A Mother's Guide

Discover Your Baby's Spirit

Is Your Child A Hero, Star, Indigo, Crystal, Or Liquid Crystal Child?

Dr. Margaret Rogers Van Coops, Ph.D

authorHOUSE®

AuthorHouse™
1663 Liberty Drive
Bloomington, IN 47403
www.authorhouse.com
Phone: 1-800-839-8640

© *2010 Dr. Margaret Rogers Van Coops, Ph.D. All rights reserved.*

No part of this book may be reproduced, stored in a retrieval system, or transmitted by any means without the written permission of the author.

First published by AuthorHouse 3/15/2010

ISBN: 978-1-4490-7743-3 (e)
ISBN: 978-1-4490-7745-7 (sc)
ISBN: 978-1-4490-7744-0 (hc)

Library of Congress Control Number: 2010900634

Printed in the United States of America
Bloomington, Indiana

This book is printed on acid-free paper.

Edited by Stephen M. Van Coops
Cover design by Teri Kahan—www.terikahandesign.com
Cover photo by Handy Widiyanto

*Is Your Child A
Hero, Star, Indigo, Crystal, Or
Liquid Crystal Child?*

Contents Page

Introduction — IX

Preface — XI

Part 1 This World and the Next — 1

Chapter 1. Our Earthly Point Of View — 3

Chapter 2. In The Beginning – Learn To Know Your Baby — 11

Chapter 3. The Five Precious Groups Of Children — 19

Chapter 4. Hero Child — 25

Chapter 5. Star Child — 37

Chapter 6. Indigo Child — 53

Chapter 7. Crystal Child — 67

Chapter 8. Liquid Crystal Child — 83

Part 2 Bringing Up Baby — 97

Chapter 9. Which Group Do You Belong To? — 99

Chapter 10. Integration Of The Five Groups — 105

Chapter 11. Your Child's First Year — 111

Chapter 12. The Second Year — 121

Chapter 13. The Next Three Years — 125

Chapter 14. Cycles Of Life — 131

Chapter 15. Awkward Growing Years — 137

Chapter 16. Watch Out For Negative Physical Indicators — 143

Chapter 17. Late Teens Into Early Adulthood — 151

Chapter 18. Religious And Spiritual Education — 157

PART 3 PARENTING THE RIGHT WAY	163
Chapter 19. Communication	*165*
PART 4 GENERAL INSIGHTS IN BRINGING UP BABY	191
Chapter 20. Creative Talents	*193*
Chapter 21. Intelligence	*201*
Chapter 22. Maternal Love	*211*
EPILOGUE	215
APPENDIX	219
In The Beginning…	*219*
The Descent of a Soul	*221*
Lower-Self Spirit Archetypes	*223*
A Spirit's Journey	*225*
The Development Of A Goal In The Soul Structure	*227*
The Development Of The Soul Structure & The Modes	*231*
The Development Of The Attitudes In The Soul Structure	*233*
Discipline Is Created By The Chief Feature	*234*
The Spiritual Centers In The Soul Structure	*235*
PRODUCTS AND SERVICES: BOOKS, KITS AND CONTACTS	237
ABOUT THE AUTHOR	243

Introduction

Your Baby Has A Soul Structure

What does that mean to you!

Everyone starts life with coding that is called a *Soul Structure*, which gives each of us our own personality and character. Everyone wants to know why they feel, think and act the way they do. Each one of us will try to understand everyone else and to model our lives on our surroundings and the behavior of others, but nevertheless, we still find ourselves to be completely individual when it comes to dealing with our spiritual awareness and the way we express it. This book is for those women and men who are now in the wake of awareness, on simple levels, to discover not only who they are spiritually, but also who their children are. To know the truth about self and others is to discover the value of the coding that is called *Soul Structure* and how it works.

Though I have written in detail about the nature and use of the *Soul Structure* in my book **The Rejection Syndrome**, it is necessary in this book, for you the reader, to understand its function and the varieties of forms it can take. It is for this reason, that I have included an informative breakdown of the *Soul Structure* at the back of this book in The Appendix. As you read, and learn with each page, you can refer to this part. In the beginning of the book, I focus more on the types of children being born and the interactions a mother and father can have. Through simple outlines and analogies I will show how this coding works in all of us and how it binds us together. I know that everyone is interested in their spiritual growth, no matter what their religion and it is my hope that this book will make your life easier and fuller with many joyful moments as parents.

Preface

This book is primarily for women and about bringing up your own baby to the best of your ability, it is also about understanding yourself as a mother and friend to your child as well as the spiritual lessons you are both learning. Once you understand how simple it is to know yourself, then you will discover it is even easier to know the many others who come in and out of your life.

PART ONE

THIS WORLD AND THE NEXT

Chapter One

Our Earthly Point Of View

Though many religions have defined what they believe to be the 'real' Heaven or Hell, it has been my personal experience through out-of-body travel and through channeling from Master Spiritual teachers to come to understand the extreme nature of The Oneness and the many levels of Ascension. The Spirit World is a complex, yet simple, state of existence in which all things are Omnipresent. Nothing is judged or discarded. All of it is a part of The Whole that we call God.

On Earth, we have learned to judge ourselves and others. We categorize things into good and bad. We use and abuse the things that are given to us. We destroy relationships, harm animals and pollute our planet. Our minds are constantly swirling with information that is so often inappropriate to our lifestyle. We place ourselves in categories of competition, societal group, class distinction, race, etc. Every place on Earth has someone who is in some way dissatisfied. If you find one who is at peace, you can be sure there will soon be a happening to take it away, or change it into something else.

Earth is a place for our spirits to come and test our evolution. How we do that is very individual. Whatever we can think of, we can do. If we harm ourselves or others, then we learn the hard way. If we take things for granted, then we lose them, and so on... It is, therefore, hard for us to consciously accept and imagine a place where none of this testing goes on. Even if we could imagine it, we would immediately destroy that image in favor of something we know.

A place of peace is a threat. It is the unknown and is, consequently, a place to fear. How would we behave? How would we exchange energy with others? What would be our purpose? Such questions fill our minds. The truth is, we cannot understand and relate to the place of peace unless we let go of earthly conditioning. We are in a conscious loop; wanting to be spiritual and to escape the pains of this world, but if we do, what will we do then?

My experiences are beyond normal. It does not mean that I am better or more improved in some psychic aspect of myself. What it means is that I have opened myself to learning and to viewing something beyond this world's consciousness. In order to do that, I had to erase the many fears that were instilled in my mind.

I learned that our *Soul Structure* interacts with our spirit's *Soul Fears*. These fears are a constant in The Oneness. They are accepted and used without pain or suffering. These spiritual fears are God's fears. We automatically honor them in the Spirit World, but on Earth, we see them as something to battle. It is our nature to survive in any way possible, and in so doing, to fight off anything or anyone, including God as we know it. God is seen as infinite. Our belief in our separation from God throughout our daily lives is unparalleled. God is too far away to hear us! Yet, God is in every face, animal or child we see; God is with us all the time.

When we die, our life ends, but our spirit continues into the Spirit World where we use our current *Soul Structure* to look at all that we have been doing during this last life. We put this life on fast forward and see everything in detail. We see the mistakes we have made and enjoy the successes too. Without judgment, we deduce our own understanding of how much we have grown and evolved in our journey to become one with God.

In The Spirit World, which we might call Heaven, every thought comes to us and through us from a basic source of pure energy. Our spiritual emotions are felt as an all encompassing energy that is constantly nurturing and supportive. If we are with other spirits, we harmonize those energies. Anyone wishing to be separated will simply vanish from our consciousness. Everything we can imagine can be manifested in The Oneness. Your little cottage with all the roses is there the moment you create it and it lasts until you become tired of it. Then it will fade back into pure energy.

All energy is in constant motion, creating friction that generates light. It is hard to think of yourself as photons of light; energy that flickers on and off as you move; yet this is your everyday earthly expression too. When you think, move and feel, you generate energy that sends thousands of sparkles of light across a room, and each one of us feels it and responds to it. Next time you feel negative, think about how much negative energy you are putting out; you are hot-wired and spark like a thunderstorm with flashes of lightning. It is a small wonder that people move away.

In 'Heaven,' there are different levels of awareness in which people congregate. Their minds and emotions become one as they support their lessons and issues. They will stay with those other entities for as long as they need to learn. Once they outgrow this plane of evolution, from their spirit self's point of view, they will move away and align with a new group. Each group will have a focus on a way of ascending and getting closer to God. They will discuss their past lives, their new ideas for spiritual growth and plan new incarnations. Some may remain behind to be helper spirit guides, angels etc., while others will incarnate as family, friends, foes etc.

Though every individual spirit has its own agenda, the integration of the group will prioritize the purpose of the incarnation. This could be to save the world from political ruin, or to bring a message of peace etc. Whatever the common goal, each will choose a *Soul Structure* that will provide a variety of ways for them to interact. Through their coding, some will use aspects of their *Archetypes to* show them how to link up on Earth, while others will use their *Modes*, *Attitudes* and *Soul Fears* to work things out. But, generally they will find one another through their *Centers* that link their spirits together. (See Appendix).

The Spirit World

In the Spirit World there are many copies of earthly schools for learning about material ways as well as the Spiritual Halls of Learning & Records.

Any spirit may attend any of these places during their stay before deciding to reincarnate. In the Halls of Learning and Records you can look at your past lives and the past lives of others. You can observe anything or anyone that has ever been in existence. The energy of those times is held and can be recreated for you to learn.

There are beautiful gardens, healing plants, wonderful musical concerts and so much more. Every spirit has a thousand and one talents that can be shared. Everyone is interested in what is being momentarily shared. If they choose to miss it, they can see it later by recreating that energy. There is no reason to be early or late. There is no reason to miss anything. In fact, reason is of the Lower Self mind and your spirit mind is full of wisdom that knows only how to be omnipresent.

While I could write reams about my Out of Body (OB) experiences and my spirit guides' tutorials, it is not my intention to wow you with this, but rather to explain to you in simple terms how you and your child have prepared to be in this life. It is too easy to believe that there was no preparation and, therefore, no link other than just having a spiritual bond. A lot of work has gone into your times together and your lessons that you are now learning. If you would like to know more about me and my OB events; perhaps reading my book *Pro-Life, Pro-Choice, Pro-Spirit* will interest you.

In *The Oneness*, you went through a *Spiritual Bonding Ceremony* before you could leave and be birthed. Some spirits bond with millions, while others bond with only hundreds. Since every spirit has gone through its own ceremonies, you can begin to understand how complex your attachments are to all those now living. "No man is an island" is a true saying. Though you may love your family for many reasons, you also love everyone else, even those you never speak to or see in physical form. Hence this is the reason you may identify with a 'super' star or a minister in some far off place, or with some ancient historic entity who lived thousands of years ago. Somehow, in some dimension, you bonded with them all.

When it is your time to be born, your mother will have prepared for you as you will prepare for your own children. Though men never have the luxury of having a physical/spiritual bonding in birthing, they can and often do still feel the same sensations their partner is having during the pregnancy and birth. This is brought on by the bonding of Spirits. It should be noted that we all live male and female lives and we all have the experience of mother and fatherhood.

So, the perception of a spirit's *Soul Structure* being activated in conception is untrue. It was long ago activated before the mother was even born. Should a pregnancy be terminated, then this has been pre-ordained too. *See my book: Pro-Life, Pro-Choice, Pro-Spirit.* An unexpected pregnancy has been planned. It is just that you do not remember making this plan.

Within the realms of the spirit world are many planes of evolution. There are dark places without much light, where friction creates more light. These limited sparks of light create more light and so the evolution is maintained with fresh energy. In the highest realms of existence, everything is an aspect of God. There is no sense of separation. Between those two extremes, all life in spirit form is being constantly pulled ever on toward absorption. Whatever your spirit mind can conceive exists within this realm. The fears you manifest will constantly motivate you and ensure the continuation of your evolution into *The Oneness*. A good analogy is that as planets in our solar system are being continually drawn towards the sun, where they will eventually be absorbed, so we will return to God. At that time God will transform. There will be a death of God as is known and a new one will begin. The energy of God never ceases to change. In a very small way, we never cease to change. Everything we do causes our energies to shift and transform us.

God's Soul Fears of Ascension

To incarnate, we must choose to work with the *Soul Fears* of *Ascension, Descension, Separation, Assimilation and Divine Love.* These are explained later in this book. The *Soul fears* accompany the Five Laws of God, generally known as the Laws of Karma, all of which have been fully explained in my other books. But, should you be a first time reader, here is a simple version:

1. Never invade another's space unless welcomed. (Separation)
2. Be responsible for everything you create in any form (Descension)
3. Integrate yourself with everyone and share without judgment (Assimilation)
4. See yourself in everyone you meet and grow spiritually (Ascension)
5. Love yourself as God loves you and become one with God (Divine Love)

To follow these Laws of Karma is our purpose and destiny. Karma simply means fate. Our fate is to return to God's bosom and to embrace the Passive God that created us. This is why we incarnate over and over again. There is no judgment, no punishment owed in Karma. That idea was manifested by priests during times when Mankind was superstitious and out of control from religious notions.

Whatever our point-of-view on anything, it is simply our way of learning something. We have so many ways to differ, but ultimately, all those ways end up in the same place. By the time we truly understand *The Oneness* and one another, we will no longer be individuals. Our *Soul Structures* will have merged and dissolved time and time again until it becomes one large one that we all share. When the Active and Passive Parts of God unite, we will have been welcomed home and once again will be content... for a short time! Then the whole cycle of creation will begin with a new theme and a new form will emerge, stimulated by one of the *Soul Fears* again. (See Appendix).

Think of yourself as an omnipotent being named God for a moment. You would not want someone to dominate you and destroy you. Nor would you want to delegate your skills and talents to someone else to misuse. You would not want to be ignored and forgotten. You would love people to honor and respect you. You certainly would like to meet people who are just like you, but excitingly different. All this is the way God has created us, a mirror of Him/Herself/Itself.

Imagine you could draw and color anything, play music of any kind, or invent any kind of instrument for your personal use. Or you could write poems and keep records of all your creations. The list can go on and on. At the same time, remember yourself in a variety of places, speaking thousands of languages with millions of people who understand your every word. Your capabilities are endless! Such is the Mind and Heart of God, and such are the levels of evolution in the Planes of the Spirit World called Heaven and Hell.

Yet, there is a destructive nature to God that is within each of us. We must finish something, and sometimes it takes a forceful way to shift energy. On Earth, flood, storms and the like cleanse the planet. We as spiritual beings need a good push from time to time to force us on. We need to let go of old beliefs and feelings in order to ascend. We can be our own downfall, and our own pick-me-up as well as that to others. In the mirror image of God, we all do this to one another throughout all time on Earth. In the Spirit World we never push. We each take our own good time to evolve, destroying our negative history, while creating a new one. Our spirit needs a constant change of energy. Our bodies need a constant replenishment of energy. These two forms of energy meet on Earth. They are called our Higher and Lower Selves.

When Life Begins

When pregnancy occurs, these *Higher & Lower Selves* must bond and unite as one. The *Soul Structures* will then harmonize. All individual Past Lives will integrate as one life's experience. Some Past Lives will be coded into the cells of the body so that talents and skills may be recovered. There is no such thing as bad karma between a mother and child. No debts are owed. Life begins in purity and harmony.

Your baby is a pure spirit being, who is wise, all knowing, God-like and ready to share. What happens after the birth will be the result of a culmination of his/her spirit's natural expression of free will, spiritual lessons and a driving need to follow their quest to satisfy their mastery of their *Spiritual Goal* which provides the ultimate lesson of this life.

Your interaction with your baby will push and pull at your own *Soul Structure* and your own spiritual lessons, while testing you through your own awareness and involvements that express your individual free will.

Being a mother is one of the greatest experiences that you can have. It takes you to the length, breadth and depth of your being. You will never be the same!

Chapter Two

In The Beginning –
Learn To Know Your Baby

"Children have a way of saying or doing the most inappropriate things at what seems to be the wrong time, no matter what their upbringing or position in life. They are by nature remarkable. In the first eight years of life, a child learns at an extraordinary rate. Through use of the five senses, sight, hearing, touch, smell and taste, a child not only acquires a great deal of physical experience, but also a great deal of silent input from within, which has been incorporated into their psyche. Your child is a child of God! They know you well!" …Archangel Haniel

Much of this book has been written and channeled by me having been inspired by Archangel Haniel who saved me in my 3rd year from imminent death, and Cornelius Tacitus (Roman Historian), together with Master Chang (Chinese ancient Spirit Guide), who have guided me in writing many of my books.

So, let us begin.

While a child is in its mother's womb, he/she must not only share a common space within the confines of the mother's body, but also her personality. During this time, the psyche of her child is connected through its own spirit to that of her own. Whatever she touches and feels, sees, hears, tastes and smells, is immediately received by the spirit of the child-to-be. In this way, genetic history is activated within the emotions of each spirit, ready to pass on emotional and mental attitudes from generation to generation within the DNA strand. This allows awareness of the mother's life experiences to be passed on to the psyche of the child. These experiences from the mother, along with inherent information retained by the spirit about to be born, are then transferred into the deep-subconscious part of the brain and on down into every cell of the child's body as it grows inside the womb. In this way, the child is born with similar emotions to his/her mother.

In every cell there are photons (amplified energy) that create tiny light beams of condensed energy encoded with negative and positive impulses that through constant friction, create more energy which when connected to the brain, forms awareness. The brain is then able to control those 'light' impulses and store them collectively in pictorial forms. Those images are then passed to and fro as the mother and child bond. By the time a child is born, he/she has all the detailed history of his/her mother, grandmother and still many more mothers in the family history of ancestors, who may be long forgotten. This 'light' energy is stored in the DNA strand which is within every living cell. This strand is still to be understood by science, but is the key to our spiritual evolution and our history which ultimately will awaken our consciousness to be truly in Oneness on Earth.

As energy passes back and forth between the mother and child-to-be, habitual ways of thinking and feeling are passed down from generation to generation. By the time a child is born, he/she already believes what is to be shared physically and emotionally as time passes. Because of this inbred habit, children never doubt their mothers until much later in life.

In various cultures around the world, women have continually justified their inbred beliefs, which have been absorbed from their mothers. Some more aware women have spent a large part of their lives trying to erase habits learned from their mothers in the hope of finding independence and freedom. Usually they fail miserably and never totally feel free until their mother dies.

Despite a need to change, or a desire to experience more, women often find it impossible to escape their history. They are continually led by conscious emotions evolved from old ones that cover their deep-seated insecurities. By harping back to the things they learned as a child, along with environmental conditions and peer groups or role models that surrounded them in the early years, they hope to find some semblance of peace and harmony. But, seemingly, no matter how hard they try to change old ideas, they are consistently drawn back to their past behavior learned in the first two years of life. Much the same can be said for men; however, they process their coding differently.

When a woman finds herself pregnant, she turns to her memories of her own childhood and subsequently rears her child according to her own experiences. In a happy home, a child is also influenced to some degree by the father and siblings, if there are any. But, despite their presence, her child will generally follow the example of his/her mother. In truth, each mother raises her child, on average, some twenty years behind the times. Hence the reason that teenagers complain about their mother and father being 'old fashioned'.

Whenever a child decides to change the historic habits of a family, this child is often seen as disruptive and inconsiderate. These children will seem to fight for their right to find their individual pathway. The more a child resists "habitual history," the more we see it as a problem that must be controlled. This often causes the child to grow up feeling resentful and unloved. Onlookers may see these children as pioneers, leading the way. Others may see a need for change within a family group, while the family does not.

Since these children may be different from the rest of the family, they may learn to grow up with aggression. Everything they do is challenged. Everything they like is opposed or denied them. It appears to them, that there is no one in their environment who supports them. As a result, they become the rebels.

Often, these rebellious children create a great deal of dysfunction within the family order. Their families rarely find harmony amongst themselves, as they are all constantly focused on this 'rebellious child.' A great deal of blame is laid upon him/her for disrupting the family's comfort. This child then feels guilty, along with developing a great need to surrender to the will of the family. Unfortunately, in a very short time, miserable emotions that have been suppressed arise in the form of insecurity. The child soon feels no sense of belonging and loses any magical joy in his life.

While all this conditioning is weighing heavily upon this child's heart, the spirit becomes emotionally weak. He begins to turn inward to find inner strength which may well outwardly show up as defiance.

During this time, the spirits are learning to create some way to be different and unique. Spiritually, children develop their psychic senses, and through them, learn to know and understand their family, teachers and friends from a different point of view. A raging mother might be seen as insecure and unstable. This child will quite naturally look for a way to stabilize their situation. Often these children will develop traits to become the 'Fixer Child.' The form and actions that these children take can vary immensely. They can be disruptive one minute and extremely helpful the next.

When children are interacting, they are processing billions of messages received in the form of photon light from a variety of sources outside of themselves. At the same time, they watch their own spirit lights, to awaken awareness of their true journeys of spiritual growth which are encoded in their deep-subconscious minds. They will try to become independent and follow their own ideas and dreams as they stimulate themselves to overcome negative physical states in their environment. At that time, they know they want to change everything, but don't know how. Confusion is the result. Millions of thoughts pass through their brain and on around their body, sending conflicting feelings and sensations outward in a variety of characteristic ways.

In order to overcome this confusion, the psyche of a child begins to develop conscious psychic skills. However, it should be noted that all psychic senses are turned on as soon as a child is in the birth process and is born. Every child sees, hears and exchanges energy with other spirit Entities who we label as Guardian Angels. Their task is to help the child to lay down solid foundation stones from a psychic point of view, which will help him/her to develop physically, emotionally and mentally into an acutely sensible and aware adult who is able to help him/herself achieve all their spiritual goals before passing back to the spirit world. Most of the negative or positive physical goals and subsequent beliefs achieved in a lifetime have been previously outlined before a child is even conceived. A spirit entering embodiment does not mind whether the life to be lived is negative or a positive one, so long as they have the opportunity to learn a lesson that will allow them to get closer to God.

The spirit of a child-to-be prepares for life long before its mother is even born. Often, generations of families prepare in the Spirit World, often known as The Astral. An evolutionary lesson can be learned by everyone over a period of time. Each generation will pass on an understanding, a way of living, with instructions to grow and develop more than those who went before them. For example: a mother, who felt enslaved by marriage, might tell her daughter to work and be independent and be careful of marrying early. She will tell her son to not enslave his wife.

During the Piscean Age, over the last two and a half thousand years, the primary focus of all children born was one of spiritual manipulation through ritual and fixed traditional beliefs. Hence there was the manifestation of a variety of religious beliefs and indoctrination which controlled Mankind in the way he or she conducted their life. Belief in God, The Creator, who controlled their lives, was strong. Those who sinned were not only cast into hell, but also cast out socially. Judgment and discrimination was rampant amongst all classes in all walks of life. Children were trained to behave and to adore and idolize their parents, no matter how unbalanced they were. They readily honored and respected their elders and followed in the footsteps of those who had gone before. Any misfits were dealt with severely, causing them to live in fear, pain, anger and guilt. Those who were 'different' were rebels without a cause - outcasts who had fought for change and never lived to see it. Those who survived were exposed to extreme measures of punishment, and in some cases, were imprisoned in an 'insane' asylum as a result of their 'pioneering work,' whether or not a person was deserving of such treatment. This was a convenient way to ignore the problem and bury the so called 'disgrace' that had been laid upon the family

In the last years of the twentieth century, we saw a big change arising. With the dawn of the prophesied 'Golden Age' manifesting in the 'Aquarian Age,' life is different. Babies being born in this time are different. Their spirits are now awakened and yearning to live in a different way from those who lived before them. No more, is a life of habit acceptable. Every spirit that yearns to be born in this time is a child that marches to a different drummer. Their vibration is different. They receive photons of energy 'light' from sources and process it in a different way. They use that 'light' energy to achieve/cause a greater shift in consciousness. Radio waves created by sound are modulated in a different way. They hear things differently and do things differently as a result.

Before a spirit is born, it must prepare for life on Earth. Each spirit chooses a coding that will allow them to live a life according to their choices made before incarnation. In a Universal sense, God is allowing all of us to become free of judgment, indoctrination, and conditioning. The result will be a unified race of humans who all see themselves as free to make their own choices without restrictions, according to their life's pathway.

For this to occur, the spirit of each child being born must return to the essence of creation. *In the beginning was the word, and the word was made manifest.* From a spiritual point of view, "the word" is "Believe." Every child is born with an intuitive sense of The Creator, and an inner wisdom of knowing. Whatever Man can think of exists. It was created long ago, and will only cease to exist when The Creator destroys it. Man can create and destroy what he makes, but he cannot destroy anything the Creator has made. Mankind can unite together as one unit and support The Creator, God, by going with the Universal flow that has its own flux. So, if God desires a change, then we as spirit entities desire change also. If the message of change is to be manifested, then what better way to do it than through children who show us clearly that change is needed now?

There have been many abominable things done by children to children of late. Our shock in seeing their ways has been a wake-up call for all of us. Children everywhere are exposed to anger, violence and war. Their perception of their parents and elders is one of love/hate and destruction. Everywhere you look, you can see what is occurring, the re-enactment of Mankind's history over the last two thousand years or more. Yes, we are shocked by what we see! No, we do not need to continue for another two and a half thousand years this way! We can change and will change because as spirit entities, we want it to change.

Like all things, change is never easy. The 'old ones,' today's creatures of habit, often have forbidden change, but those who follow them now, invite change and are starting to get the ball rolling. But, it is the future generations to come that will finally herald the changes with great joy, embracing a new kind of life, where independence and unity is the norm.

Chapter Three

The Five Precious Groups Of Children

Today, there are five very different groups of children being born. Each of these groups brings a lesson of physical, emotional and mental change to those who interact with them. They are our wake-up call. Many of these children are themselves emotionally, mentally and spiritually unbalanced. Their perception of the world is warped and twisted away from the norm. They are vibrant, independent, strong and stubborn. Their brains are over-active, acutely psychic and extremely clever. Their inner knowing is without doubt. Their lust for freedom will stimulate them to ignore indoctrination, rules and regulations. In short they are a different breed of Mankind.

How then can one help these children find their niche in this life? Certainly not by quoting family history, rules and regulations! Their way is by accepting the use of *The Oneness*, which is the evolving form in which The Creator lives. In The Creator's world, all is manifested and all is perfect. There is no imperfection! Only constant change! These children are the change! They can settle for nothing less than change.

The first group is anatomically different. They are the *Hero Children* who carry a new strand of DNA. This group of children is stronger, bigger, with a greater capacity for intellectual change. Their perception of the way they see life is practical and easily changed according to their desire. They quite simply watch, listen, weigh up the situation and then know what to do to bring about a change. In a dysfunctional family, if someone is causing pain, they will eliminate the cause. In a functional family, they will stimulate those around them to change for the better. Whether their influence is negative or positive, they are the physical leaders of the future. They are the re-organizers who will establish a new perception of living. They are the 'Spiritual Hero' Children.

The second group, are children of the past. They are the *Star Children*. Their sole purpose is to assimilate what has gone before. They are very normal and easily led. They require a great deal of attention and are full of questions that need satisfying answers. They will settle for nothing less than the truth. Their entire life will be spent searching for answers to the truth. They can be lost in illusions and fantasy or grounded in spirituality while in observation of themselves and others. These children will become the men and women of the future who will effectively fight for change, so that the other four groups can function. They are the amalgamators of things past with things to come. They are the clean-up crew. They bring a lighter consciousness for a better future.

The third group, are called The *Blue/Indigo babies*. These children are God-like. They see only what is right emotionally. They are acutely sensitive to their surroundings and will influence those they meet to be child-like. They are the mediators between God and Man. When negative, they are capable of emotional tantrums, disruptive actions and inconceivable behavior. However, when positive, they are capable of producing great works of art, talents beyond belief, and they express emotional insights from beyond this world.

The fourth group stimulates us by being Creators. They are the *Crystal Children* who love the elements of the Earth. Their intention is to have everything new. They are highly intelligent and very emotional. They are impatient and often unable to focus on what was or is happening. They constantly search for new interests, becoming easily bored with people, toys etc. once they have mastered them. They are experts in being center stage, and will go out of their way to gain attention. Their psychic senses are strong as is their spirit of adventure. They love to express their talents in the arts and to share their skills in a competitive fashion. They are the sportsmen, the dancers, the musicians, the artists, etc. Their very existence is the manifestation of change.

The fifth group, are born spiritually aware. They are the *Liquid Crystal Children*. Their uncanny ability to talk to the spirit of a person is mind blowing. They know what you are thinking and feeling long before you can voice it yourself. They know all about you. They are exceptionally psychic. They commune with spirit Guides and God as normally and naturally as speaking to their mothers. They see only truth and hate liars. Their life is lived for truth and those that are bombarded with negativity will withdraw from the everyday world. They often live in so called 'fantasy', speaking in tongues, expressing talents where none were shown before. Their spirit is spiritual, their flesh is fine like their spirit; they are spirit in expression and easily destroyed. Those that are nurtured are a delight. They speak of the joys of life and live to play. They are delighted by everyone and are without judgment.

In all five of these groups, there are some children who need special care. Many of them are dyslexic, ADD, ADDH, Obsessive Compulsive (OC), neurotic behavior, Tourette's Syndrome, Autistic, Single-minded, aggressive, activists, and more. In each case, their spirit has not easily accepted their environment and the life they have chosen to live in. They are in discord - lost between the Universe of the Spirit World and this. Their very presence on Earth is to awaken us to the changes they bring into our lives. They are all dreamers of things to come and, as such, must be helped to bring their dreams along with their personalities into embodiment, so they can function as we expect normal children to do.

These children need adults to become one with them in their world. They need parents who are aware of their needs and able to see and hear through their eyes. These children perceive all things though their five physical senses as well as through their five spirit senses. These combined senses make this type of child acutely sensitive to their parents. Therefore, the parents should be open and honest at all times, even when negative events occur.

All children believe in miracles, magic and the joy of life. It is important that parents find this belief within themselves. Old beliefs should not be attached to these children. It they are imposed, they will then only rebel. Parents must be able to give solid support at all times. Vocal communication is important. The voice must be lyrical and melodious. Sharp tones admonish and destroy. There should be no assumptions made about these children. No matter how dysfunctional a child may appear, it must be accepted that this child is demonstrating something that needs to be seen and learned and is, therefore, happily expressing a truth. Remember, "Out of the mouths of babes, comes wisdom."

Children that appear 'normal,' are perhaps under greater stress and tension at this time. Their psyche is active and their awareness is uppermost in their minds and emotions. Life is not easy for them. They are not able to obey without question. Their sensitivity to everyone can be overwhelming. If these children are exposed to a great deal of negativity, then their spirit is often broken, and their character and personality destroyed. It is important to encourage these 'normal' children to do well and to feel wanted and accepted. If they are pushed against their will, they will develop 'passive aggressive' natures.

Every child is complex. There is the coding of his/her own spirit, mother's spirit and personality and ancestral DNA, along with training in the first years of life. During that time the child is in hypnosis. Everything the parents, teachers, family and friends say is absorbed into the sub-conscious mind. There is no critical mind to protect a child from hocus pocus information. They believe everything. Only the inner spirit mind, called the deep-subconscious mind, is able to distinguish between wanted and unwanted information. In these early years, that unwanted information is cast out, and in its place is a lesson received. For example, if someone tells a lie about fairies and says they do not exist, the spirit of the child knows that they do, and so casts out the negative suggestion.

Later in life when rationalization occurs, once the critical mind arises around eight to ten years old, a child will accept false information, which can contaminate the sub-conscious with direct inhibition placed upon the deep-subconscious. The end result is a lack of belief in oneself and the ability to do as one desires. Children grow up discounting themselves and others. They lack imagination and creativity and soon find themselves lost in pain and suffering.

These new groups of children will not allow this suffering to continue. They prepare us now for future generations to come, where there will be no lying, cheating or pretending. The psyche will be well developed and communication will be done through the psyche. But until such times when this does occur, we must deal with these children who seem to be beyond our reach or understanding at times.

Hypnosis is an altered state of consciousness in which all children live until they are approximately eight years old. In an awakened state of hypnosis, a tremendous amount of information is absorbed, hence the ability to learn so quickly. In this state of awareness there are great moments of joy when learning is at its highest. During those moments, the vocal and emotional content of the experience is highly suggestive and easily absorbed. Story telling is exciting and enjoyable. Mothers love to act out the parts and explore the sounds and tone of those involved in the story. The child learns to identify positively with those sounds. Primary programming demands tonality.

To help a child to find his/her footing in this world, these tones and pictorial images must be encouraged and explored. Added senses, such as smells and feelings, along with touch, bring the lesson home. Children who are potty trained by expressive mothers soon become clean. The more physical the mother is in her actions with the child, the more the child learns. Her dialogue should always be encouraging and supportive, even when the child makes a mistake or does something wrong.

While every mother is in some way supportive to her child, there are always those times when in an unconscious moment, the child is set aside in favor of another child or some pressing situation. During those moments a child will process their feelings, along with their other senses, which will make a marked impression on that child for the rest of their life. Such an example could be, that mother is on the telephone speaking about an important issue, when her child runs up to share the news that he has seen something wonderful on TV. She tells him to be quiet while she is talking and, in that moment, the child is abandoned and his experience dismissed. He then creates a belief that he cannot share himself with others.

It is an accepted fact that no parent can be aware of a child's total being, emotionally, mentally, spiritually and of course physically, all the time. Bringing up a child is a long and arduous job that requires a great deal of love, trust and patience. Throughout the years of the growth of a child, there will always be challenges and opportunities that provide instant moments of awareness and experience. Each child will process its own experiences differently. No two children are the same. Within a family, the siblings will perceive events differently and according to their personal point-of-view, teaching themselves how to think and feel about many issues.

Over a period of sixteen years a child learns to think for him/herself and does not want to be told what to do. Teenage years are often the hardest to deal with from a parent's point of view. Often rebellion and disorderly conduct can result.

No matter what the circumstance or the events that occur around a teenager, there is always an inner sense of a need to belong to a group. This desire to fit in and belong stems from their spirit consciousness. Every generation of children has a mutual focus from a spiritual point of view. The outer planets of our solar system bombard us with photons (electromagnetic particles of light that emit through a vacuum) of energy that stimulate the collective consciousness of each generation. Daily, we each strive to interact, share and grow together according to our age group. Today, teenagers are being prepared to shift this Earth into a new gear. We have destroyed and now we must rebuild.

So, this book is about understanding the kind of child you are rearing and how to help him/her to become the best that they can possibly be.

Chapter Four

Hero Child

Children With An Extra DNA Strand

When these children are born, all their psychic senses are active. They easily attune to their surroundings and quickly find their place in society.

Through the use of their psychic senses, they develop awareness at an extremely early age. In fact, within the first hours of birth, they are spiritually conscious of the effect and the use of energy that their presence creates. They will experiment, e.g., cry to watch the parent's reactions.

An extra DNA strand causes these children to grow faster and to develop a higher than normal intellect, as well as a more healthy bio-chemical balance. By the time they are three years old; they can read and understand further advanced levels of education. You might say that they are geniuses compared with children born 50 years ago.

However, this enormous capacity for learning comes at a price. Each child will have selected a coding that was programmed into their spirit, which in turn, was programmed into their physical body. This primary coding is called a *Soul Structure*. (See Appendix).

Every child is born with a *Soul Structure*. However, these special children with an extra DNA strand have five aspects of the *Soul Structure* in common with their type. Their coding is very similar, yet totally different, which ensures individuality.

The first aspect in common is that they are all programmed with a *Mode of Power*. (See Appendix). This gives each child a personality which will develop a need to either be giving in leadership, (positive) or tyrannical in command (negative). Whether using positive or negative aspects of this *Mode of Power*, it always creates a person with a desire to rule.

The second aspect within their coding of the *Soul Structure* is a *Goal of Growth*. (See Appendix). This means that they are motivated to explore and change those things which are not acceptable to them, and which they feel they can control. A group of these children can carry out an act that will change the world in some way.

The third aspect they share in common in their *Soul Structure* is the spiritual coding of the *Higher Spiritual Emotional Center* (See Appendix). This gives them a strong need to correct the flaws of his/her world. By relating intelligently in an analytical way to a basic issue, he/she can bring insights from their Spirit-Self into consciousness. In this way, they accept and understand a need for change and will make the changes as soon as is humanly possible, without procrastination.

The fourth aspect of a common individual coding in the *Soul Structure* is a negative trait, called a *Chief Feature*. (See Appendix). This trait is known as *Martyrdom*. If they are failing, they will sacrifice themselves for the common good of humanity. Therefore, each child needs a 'cause' to justify his/her existence. They may become doctors, researching new cures, or politicians finding new ways to harmonize nations in trade etc.

Whatever, their choice of work, they will always yearn for supportive organization and constant improvement in circumstances. They will not accept entrapment through old history repeating itself, or people in a state of 'helplessness.' They can be ruthless, but loving and kind too. These children will be whatever the world needs them to be to change the world for the better. They may do something really bad to teach the world a lesson.

The fifth aspect in the *Soul Structure* is their *Attitude of Idealist*. (See Appendix). With this trait, each child has a need for perfection. They live to expect all things to be or become perfect. If imperfection surrounds this child, then he will grow impatient, angry and disillusioned. This child is hard to raise, but with the following guidelines, they can become wonderful individuals who will do great things.

A Hero Child's Influence During Pregnancy

Realizing that scientific research looks for analytical proof of the nature of an unborn child prior to birth, much is still mostly unknown. However, from a metaphysical point-of-view, there is some information to share.

Before and during a pregnancy, the mother's spirit will integrate with the spirit of her unborn child. She will accept the coding of her child's *Soul Structure* along with his/her extra DNA strand. Her body will then adapt and, as a result, will transform her ideas to change her outlook on life. She will think and act differently in the way she does things. In other words, her personality continually changes.

She will become more demonstrative in her quest for the ideal. If she is primarily negative owing to a bad history, then she will be quick to try and change her ways. If problems are tying her down, such as a bad marriage or job, then she will remove herself from that influence. She will have a strong sense of capability to rear the child alone, even though her material world may be unstable. Her independence will slowly emerge and by the time the child is born, she will have strong beliefs and understandings about her role in her child's life. Despite the hardships, she will fight to keep her child. Her role as a leader for the child will become acute.

If she has had a positive upbringing, then she will be planning a bright future for her child, being conscious of a need to find good schools, etc., even before the baby is born. She is likely to be interested in speaking and teaching her child while he/she is still in the womb. If this is the case, the child will learn and be born with knowledge that will be expressed once the child has autonomy and can speak.

In the case of each child born with an extra DNA strand, the mother's way of life will be extremely influential in the way her child processes his/her experiences. In time, he/she will join with others of like ideas. Together they will bond and carry out various acts, which may seem destructive. But, the end result justifies the means. The results they long for are a positive and happy world.

How To Bond With Your Hero Child In Pregnancy

Your basic intuition will help you notice how you are changing. If you become overly obsessive about things being perfect, then know you have a child with an extra DNA strand.

Talk aloud and inwardly to your child in the womb. Your voice will be heard. Be strong and use loving tones. Talk positively about your life to those who are in your life. Even when things are going wrong, search for the positive aspects and emphasize them. If there are major problems, talk them through aloud, and resolve each problem practically, emotionally, mentally and spiritually. Remember, your unborn child has insights already and can inspire you with solutions.

Give your child physical stimulation, by rubbing, massaging the abdomen daily. Send loving feelings from your heart through your body to your baby. Tell your baby how perfect and wonderful he/she is.

Daily plan a new experience for you to explore. Learn and enjoy any opportunity that will enable you to grow and expand your consciousness. Be the example. In this way your child will be born with confidence.

Face fears and phobias. Talk through each episode with a counselor. Your child will hear problem solving and be born with an ability to overcome obstacles.

Be physical; try swimming, climbing, riding etc. The more physically active you are, the more your baby will be born with a passion to also be physically active.

Spend time with those you love and build trust and confidence in all your relationships. Your child will be born with an innate sense of trust.

During the pregnancy, your baby will give you his/her unconditional love and support that flows from within his/her *Soul Structure*. Your energies will be high and you are likely to be healthier than you have ever been. Your sense of well-being and love of life will be strong; while your emotions may be overly sentimental as you process your life.

All in all, as you and your baby entwine your spirits together, exchanging energy, you become one, and by the time you give birth, you and your child will be ready to begin a journey of growth and discovery.

The Birth Of Your Hero Child

A child with an extra DNA strand should be born naturally whenever possible. He/she needs to begin life by entering through struggle to stimulate and activate the coding in the *Soul Structure*. As motherhood approaches, you should be focused on natural childbirth and should do everything in your power to relax and go with the flow of birth pangs.

Once your child is birthed, you should hold him/her for several minutes before anyone cleanses and washes your baby down. This is an important moment allowing the spirit of your baby to bond physically with you by using their psychic sense of psychometry. Then where possible, the second person to hold your child should be his/her father, or male relative. This gives your child a connection to the female/male energies of his/her existence. Only after this has occurred should the doctors and nurses take your child.

If this is not possible, then your baby will bond with the first caretaker who takes over and, as this is a different source of energy, your child will activate a need to resist. Later in life, your child will likely resist parental control completely. If bonding is allowed immediately after birth, then your baby will honor and respect both parents and be stimulated to enquire and learn from you and your man/husband. Should you be forced into unusual circumstances where you are not able to bond with your baby, then as soon as you can, spend several hours just holding, speaking and reassuring your baby to accept that you are his/her parent. Let love flow.

Special note about Hero Children

It is interesting to note that children who resist their parents later in life were activated by circumstances at the time of birth to believe he/she is in abandonment and consequently will spend their entire life seeking acceptance. While neither of the above is right or wrong from a spiritual point-of-view, I sincerely suggest that it is always better to seek the lighter more positive approach for all concerned.

If this Hero Child is born into a negative environment, then he/she will constantly feel separate, alone and disillusioned with life. He/she can develop abnormal behavioral patterns that may need psychological and medical help.

This group of children can develop Attention Deficit Disorders (ADD or ADDH) or passive aggressive ways making them very disruptive and unmanageable. This is their way of asking for an opportunity for change. In their world everything is new and important. Nothing is lasting or constant. If they are hyperactive, they need constant physical activities that teach how to move energy constructively. If this type of child is left alone, they can be self-destructive.

It must always be remembered that any pregnancy can only occur when the spirit of the mother is in harmony with the spirit that is to be her child. Their lessons will be identical, though their perceptions of those lessons may be very different. It is important for a mother to constantly question her beliefs, state of mind and emotions and her resulting activities. If there is a tremendous amount of fear, pain, anger and guilt present, then her child will be born with these traits.

When an extra strand DNA child is exposed to negative traits, this child will become over reactive to all stimuli, which may result in his/her doing something completely unacceptable; i.e. attempting to commit suicide or hurting another person. This is their way of crying out for help and in obtaining acknowledgement of their needs.

Tender Moments after birth with your Hero Child

In the first moments of birth, touch your baby and speak calmly in a low voice to your Hero Child. Tell him/her how much you love them. Let your child feel the spiritual bond between you. Sleep with your baby in your arms for a while and then give your child a new experience of being alone in the crib. Speak to your baby and reassure him/her from across the room. This physical separation act is important. The Extra DNA child needs to learn immediately that independence is in order. Usually, this type of child will adapt and sleep well alone.

Breast feed your child, if you are able, for a short period in time and wean as soon as possible. Your baby needs to learn separation right from the beginning, which strengthens independence.

When bathing your baby, sing to him/her to distract their mind from the insecurities of cold and exposed feelings. This will teach your baby to enjoy her/his nakedness and to release any fears of vulnerability.

Constantly talk in reassuring tones and tell your child how loved and good he/she is. Remember this Hero Child needs perfection. But, don't spoil him/her by making it too easy. This type of child needs to stretch their mind.

These early days following the birth will stimulate his/her *Soul Structure* to focus on becoming naturally independent, strong and capable. These simple techniques in caring for your baby will strengthen their character no end.

Hero Babies Have Soul Fears

Every child is born with one or more *Soul Fears*. These fears are a primary stimulation that pushes a child to mature and live a full life. In a spiritual sense, these fears make sure that each individual is reminded of their connection with God and to feel and accept inner peace with *The Oneness* (All that has been created). While it is possible for a child to have more than one *Soul Fear*, only Ancient Spirits can deal with all five *Soul Fears* being encoded into the *Soul Structure*.

The first: *Soul Fear of Ascension* is a fear of spiritually evolving to become a part of God. In physical form, it is a fear of success. If a child in this Hero Group has this Soul Fear, then he/she will procrastinate in their leadership, but be driven to overcome this fear.

The second: *Soul Fear of Descension* is a fear of spiritually falling from grace. In physical form, it is a fear of failure. Children in this extra DNA group with this Soul Fear will develop a deep need to be accepted as a leader. This will stimulate them to try harder to prove themselves worthy of leadership.

The third: *Soul Fear of Separation* is a fear of being cast out in independence from a state of God consciousness; in all senses to become Godly or God-like alone. For a child in this group, to stand alone and unknown is a requisite to greatness as a leader or performer. They will not accept anything but the best.

The fourth: *Soul Fear of Assimilation* which is spiritually the fear of losing one's identity in the evolution of absorption back into *The Oneness*. For a child in this group, this fear of not being popular and standing out in the crowd, will lead to the development of a strong will in leadership to get their ideas across to others. Denial of leadership can be devastating and lack of support can create delusions of grandeur. This will cause a neglected child to act in ways that will gain more attention.

The fifth: *Soul Fear of Unconditional Love* is the ability of the spirit to exist in acceptance of all that is created without judgment. In physical form, this is mirrored into the child's attitude. For a child in the Hero group, if the mind is full of negative beliefs that cause suffering, then he/she will form an opinion and act upon it. This may be to cause great harm or the reverse. But, whatever they do, nothing will remain the same. In this way, they will make changes on the Earth, which in the long run will result in a better place to live.

If a child's spirit is 'old or ancient' and is dealing with more than one Soul Fear, then a child in this group will become a spiritual mentor/guru or commander/politician. Either choice demands attention, and gives them an opportunity to lead, teach their lessons and beliefs through their own life's existence.

Whatever the coding of your child's *Soul Structure* is, if he/she is in this group, you will find that you have a child who is emotionally deep and often disturbed by what they see and hear. It is important that you encourage him/her to communicate their feelings and thoughts. With good guidance, he/she will mature to become a fantastic adult with many talents and skills.

Support Your Hero Child Positively by:

- Physically touching your child while showing him/her something new
- Always talking in positive sentences
- Explaining your actions simply
- Allowing your child to explain and express their feelings and thoughts at the moment of experience

- Expressing importance and understanding at the time of an event or happening
- Supporting their changes in attitude
- Be accepting and listen to their ways of expression. But, if they are negative, being firm without compromise to teach focus and intention
- Encouraging emotional exercises to develop laughter and song
- Encouraging the belief in positive manifestations of desires
- Providing organized, but flexible routines.
- Developing structure/routines along with habits that develop confidence
- Discerning behavioral negative habits and helping your child replace them with alternatives of their choice
- Always talking problems through to a solution
- Encouraging independence with praise
- Helping to develop a good attitude about Mind/Body/Spirit
- Teaching harmony in emotional and mental concepts
- Helping to develop skills and talents
- Being a good role model and acting as you speak.
- Encouraging them in their interests in the paranormal, psychological pursuits

Protect your Hero Child From Negativity By:

- Refraining from speaking harshly, yelling, screaming etc.
- Avoiding accusations, blaming or shaming them, especially in the company of others
- Avoiding giving punishment with violence or denying their skills and talents

- Not keeping your child from socializing
- Avoiding ignoring his/her feelings and their need for discussion
- Refraining from speaking for them
- Not choosing for them
- Never blocking creative imagination, expression and actions that develop their inner awareness
- Never denying their reality
- Avoiding leaving your child without leadership support and training
- Avoiding training with inflexible, ridged control and unnecessary routines
- Never dominating or ordering this type of child without discussion first
- Not denying their loyalty to family and friends
- Never telling them him/her that they know nothing
- Always avoiding turning your back or becoming moody with your child
- Preventing anyone from blocking their ability to learn
- Refusing to constantly do things for them
- Refraining from telling them that they are stupid, backward, or lacking some ability
- Refraining from lying to them
- Not pretending something is different from their perception of what is actually happening
- Not denying their psychic ability

Chapter Five

Star Child

Children of the Past

While it is not the best title for these children, since they themselves think of their future and the world they long to see, it is from the writer's point-of-view, the best way to describe them.

They are like any other child born in the past century, with a desire to see a better world in which to live in. Their focus is down to earth and practical, with many rigid ideas and disciplines. Yet they desire to reach for the stars! These children are the men and women of the future who will clean up history and make it sound good. They will rationalize, excuse and explain away the past as they prepare to make better changes in the future. They will build upon what was good and throw out what was bad.

These Spirits now being born as Star Children have chosen to follow in their forefathers' footsteps. They are pre-programmed through genetic coding to think and feel in the old ways, even back to the beginning of time. In a sense, they are innate history buffs. Their personal quest is to get things done the right way.

Each one of these children has many past lives encoded into their DNA strands. They are rather like lemmings, who act without really knowing why, yet deep inside they have this longing to find a new reality, one that will make them feel safe.

Since there is no such thing as 'bad Karma' as western philosophy would have us believe, there should then only be 'good Karma.' But what exactly, does Karma mean in the eyes of these bright new children? In their world, it simply means 'fate or destiny.' In this 'New Age of Consciousness' their destiny is already pre-ordained.

Children in this group are born with a clear purpose to simply not repeat history. Their past lives are encoded with all the positive lessons they have learned as both leaders and followers. Their innate memories will stimulate them to overcome obstacles and fashion new ways to function, with as little danger and destruction as possible. In truth, they are the garbage cleaners who will waste no time in cleansing the world of negativity.

Collectively and individually, their encoded memories will awaken them to leadership and the dynamics of change. Those who are to become supporters and followers will also prepare in like manner. Together as an organized body of Mankind, they can effectively make changes. Whether they go to war, or create a spiritual union is yet to be seen. But, whatever they do, they will act with speed and our Earth will never be the same after they have finished their work.

Special Note About Star Children

Like all things in life, the beginning of life itself is innocent and vulnerable to attack. These children in this group are perhaps more vulnerable than any of the other groups. They expect more of themselves and others. They will want 110% effort from those they work with and will give as much themselves. They will be the workaholics, the organizers, the arrangers, the politicians, The Police Force and so on. God's Laws are to be followed by everyone, but these children are those laws made manifest. Yes, they are encoded with God's laws as though they were written upon their forehead.

Though these laws are written in old style in my other books, I have modernized this version for you.

The first law is "No person can invade the will and space of another at any time." Therefore, this type of child, when threatened, will suffer no qualms about punishing the wrongdoer.

The second law is "Each person must be responsible for their actions and the results of those actions." So, this type of child will be quick to reprimand and teach those who create misery in his/her life or the lives of others.

The third law is "Each person will love others unconditionally as they would expect to be loved." Children from this group will be outspoken about philosophy and structures of education where moral and ethic codes are broken. They will seek fulfilling love for themselves first and then share it with others.

The fourth law is, "Don't judge the people around you, because like attracts like in the mirror image." What goes around comes around and these children will see you as others see you and adapt without judgment. It is therefore, important for each of us to see ourselves as others see us. In this way these children will see themselves also.

The fifth law is "Each person must surrender to the will and love of God by expressing unconditional love in order to receive all that is good." The children in this group will be conscious of good and bad, and will establish new parameters for humanitarian ways to evolve.

This group of children will all have a long and hard journey through life, but they are well prepared. Just like the bees in a hive who work diligently to serve the queen bee and the wellbeing of the hive, so each child works to honor and respect God and all that has been created for the betterment of Mankind.

When these children are born, their spiritual coding of the *Soul Structure* is hidden deep within the recesses of the mind. Their psychic senses are integrated into their everyday physical experiences. For all intent and purpose they appear to be very normal. But, there is one difference: their minds are strong and once they learn to harmonize the power of the mind, anything can manifest. Their motto should be "You are what you think!" Every Star Child is sensitive and sensible, yet oddly strange in their own way. Their will is strong and their desire to be in control can never be daunted, even in the worst of situations. They will always find an alternative way to find happiness and success despite resistance.

A Star Child's Influence During Pregnancy

Usually, the mothers of these children are planners. They have followed a dream of becoming married and having achieved that goal, immediately begin planning a family with a nice little house, car etc. Despite their emotional planning, they may conceive at a time that seems to be most inconvenient.

Generally, a mother's health will be good, but her mind and emotions may well be out of balance. Her attitude to life is likely to be one of a *Realist*, (See Appendix), whether it is coded into her *Soul Structure* or not. Her nature will be to act upon information and come to conclusions that may be negative or protective of herself and her lifestyle. In extreme cases, she can be very negative and bitter.

Because she is already programmed to believe in her past, her planning for a child will be essentially to fulfill herself in the quest to procreate. She will in many ways, be similar to her mother in her looks and emotional/mental attitude. This copycat history is perfect for the spirit of this type of child to be born.

Though the mother may be fighting herself and her history, by claiming to be changing and subsequently trying to break away from her own mother, she will in fact continue to repeat history, by simply demonstrating it in a new form, unaware that she is repeating an emotional pattern.

While the spirit of her child is in the womb, her DNA structure becomes encoded with all the history of her forefathers, especially the emotional lessons of independency that have been passed through the long line of mothers from times gone by. As the genes divide and form, this coding will cause anatomical throwbacks in the development of the baby. For example, a blond couple could produce a dark haired baby just like great great-grandpa. Or, some ancestor could have been oriental and suddenly the child is born with doe eye shapes.

There is no pre-defining the ultimate conditioning of the DNA in these types of children in this group. They are naturally drawn to the earth, sea and air. Their bodies can be carrying past genetic diseases, deformities, allergies and emotional and mental imbalances. They can also be carrying the genes of the heroes and geniuses, along with wonderful inherited talents and skills.

As a result of all this history, the mother of such a child in this group will be highly concerned that her child is growing well in her womb. She will constantly worry about her health and watch herself for signs of anything going wrong. She is most likely to suffer with nausea, and early womb pains in the first trimester. Or, she may have troubles, such as blood disorders or swelling of the ankles towards the end of the pregnancy. She is likely to become overweight and feel burdened by the pregnancy and will long for the birth to take place.

It is important for a mother carrying this type of child to nurture her body with massages, warm baths, good food, and plenty of rest. The spirit of her baby-to-be will absorb her personal life's history during the first few hours of ovulation from her Spirit. This is done outside of the body where a pure connection of Oneness can ensure a spirit-to-spirit harmony. This spirit will also oversee the growth of the baby in the first trimester, during which time his/her new mother will experience out of body (OB) events often. These OB experiences are then dreamed out into conscious awareness upon awakening. These OB events also allow the child's spirit to observe and connect with their Spirit Guides in readiness for life. During this time, the child's spirit is still partially separate from his/her mother, which is why she feels some sense of fear in the pregnancy not being perfect.

Once the second trimester is entered, the spirit of her unborn child will completely bond with her, giving her a lift in vibration. This is caused by the presence of the Spirit Guides of the baby-to-be who are now bonded with both the spirit of the child and mother as well as her own Spirit Guides. This is actually a set of four different relationships happening at once. At this point, a karmic sense of belonging arises. The pregnancy becomes the all important thing in her life. The Spiritual Union is complete.

If she comes from a dysfunctional family, she will subconsciously lay great hopes on this child being the 'fulfillment of her life' to comfort her and make her life whole and safe. This neediness will stimulate her child to be born with a need to find love and a quest to understand his/her role in life.

If the mother is content and has had a balanced upbringing, then her need to give her experiences to her child will be uppermost in her mind. She will want to educate and plan every aspect of her child's life. She will dream of a perfect relationship with her child. However, her child will be born with emotional resistance in order to further develop a growth in independence with a desire to become involved in a quest for new experiences.

While it is hard for a mother of a child in this group to be objective about her child's independence from her, it will help a great deal if she straightens out her lifestyle and establishes a routine that will give her child good guidelines to follow in preparation for a solid future. In this way, her child will respect her and his/her elders and peers. It will be important for him/her to be able to establish a good career that will open doors towards success. The joining of many of these children from this group will effectively preserve Mankind.

How To Bond With Your Star Child In Pregnancy

The spiritual connection between you and your baby is very important, so in the first trimester it is essential to have time out to meditate and focus on acceptance of your child's spirit and then to harmonize with the coding of your child's *Soul Structure*. Any dislikes, hatreds or negative traits that you have should be ironed out before the birth. You can do this by consciously accepting insights that arise from intuitive moments of awareness. These will guide and direct daily activities in positive ways.

Joy and pleasure should be your focus during this pregnancy. Spending time planning your baby's room and filling it with all the things he/she will need, will develop an emotional bond between you and your child. It is important for a Star Child to feel welcomed into the world. Remember that he/she will be listening and absorbing all your thoughts and feelings in preparation for life, so good hearted discussions about the baby are important.

You are likely to sense the time of pregnancy drawing close, as you think about the idea of having a baby. In the weeks before ovulation, your psyche may sense a spiritual entity nearby. Yes, it is likely to be the spirit of your child-to-be or the Spirit Guides of him/her to make sure all goes well. Your intuition will tell you when you have conceived.

Once you know you are pregnant for sure, begin emotionally talking to your child by speaking internally about your things of interest. Describe the things you like to do and then do them. This will help him/her to develop those skills later.

Communicate expressively about your life along your hopes and dreams to loved ones and friends. Include the baby in these conversations. You can be sure your child is listening.

Place your hands on your belly and send energy from them into your abdomen to connect your life force with the child's. When you do this, you will get a physical reaction, such as a twitter, nudge or kick, depending on how advanced the pregnancy is.

Read books that are informative and develop your intuition to awaken your Star Child's psychic sensitivity. Your baby will use these senses once he/she is born.

Be positive about yourself and your life and face any fears of failure. Let your child know you are a problem solver and happy to explore changes. In this way, your baby will develop those traits too.

Remember that your child will 'mirror' you and your behavioral traits, so be sure to lay a good foundation for him/her by being a good example, even before birth.

The Birth Of Your Star Child

There is no telling which way is up for your baby, and he/she may come feet first; or perhaps, arrive sooner or later than expected. Even caesarean births and forceps delivery, fall into this category for this kind of child. Strange places, such as ambulances, taxis and hotels could be the delivery room, so make sure you have someone reliable around to help you when the time does come.

Yes, your baby will enter this world any way they can, and as soon as they can. So, it is important to breathe and relax and go with the flow of the contractions. Any tightening of the muscles can cause your delivery to be prolonged and that could stress your baby's heart.

Once your child is born, and the cord is cut, you should hold your baby first before any cleansing activities occur. This will stimulate the core of the baby's *Soul Structure's* coding to kick into play. Then it will be safe to pass your child around to family and hospital workers. As this happens, he/she will quite naturally accept the many feelings received from different energies in the room. He/she will be busy comparing everything with past associations from his/her previous lives, though you of course can't see that happening. Within hours of birth, your child will accept its arrival and will adapt reasonably well to changes as you make them, according to his/her needs.

Tender Moments After The Birth With Your Star Child

When you hold your baby for the first time, touch his/her hands and feet and then his/her head and face. These sensory actions between mother and child will stimulate his/her brain to become aware of his/her physical body. This simple method of touching stimulates your child's brain from a spiritual point-of-view to start the history tapes running. Your baby will associate all sensory experiences as 'button pushers' into a library of history where insight and inspiration will be stimulated to manifest within his/her brain as pictorial images waiting for later explanations and understanding.

Speak softly and sing to your child. His/her auditory senses are acute. If you make abrupt noises, he/she is likely to be frightened. Early childhood fears can be developed from such noisy shocks. So, tread and talk softly. This Star Child is controlled by what he/she hears, rather than what is seen.

In the days that follow, praise your child with direct statements such as, "You are a good boy!" "You are a beautiful child." Positive programming is important. Your child will build his/her personality according to these early impressions of how the words and your personal world sound.

Star Babies Have Several Soul Fears

Children in this group may have a single or variety of *Soul Fears,* including as many as all five. (See Appendix). This mixture of *Soul Fears* is relative to their spiritual lesson and the spiritual age of the child. Below are the outlines of a spirit's Soul understanding in the evolutions of *The Oneness.* However, it should be accepted that there is no reality, rule or guideline that is fixed. Every spirit is free to ascend in consciousness in its own good time. There can never be a forcing of spiritual growth. We each chose our own *Soul Fears* as made clear below.

If a child is a 'Young Soul Spirit,' then they will most likely have the *Soul Fear of Unconditional Love.* These children will be vey trusting and accepting of the ways of the world. Their habits will be easily formed and they will try to please everyone. But, as life evolves, and the rest of their *Soul Structure* kicks in, they will develop an inner strength to overcome the negativity of the world after many hardships. Should they have chosen another one of the *Soul Fears* to work with, then their character will develop around it accordingly.

If the Star Child is a 'Mature Soul Spirit' they are most likely to have chosen two or three of the *Soul Fears.* These fears will challenge them during their life to overcome the need to be supported and approved. In time they will learn that self-importance is the best way for them to acknowledge their lifestyle. By being true to themselves this way, they can achieve the best results. The combination of chosen *Soul Fears* may create a sad or happy journey.

If a child is an 'Old Soul Spirit' they are likely to have three or four *Soul Fears* encoded into their *Soul Structure.* Their journey is one of learning to erase *self-importance*, and replacing it with *self-realization* in acceptance of the total self. (See Appendix). The total self is the state of being both in harmony and acceptance of the physical, emotional, mental and spiritual self that is in Oneness with God. This child's journey is likely to be one of suffering, with turmoil after turmoil occurring in the beginning of their life. Later, this results in a revolution somewhere in the mid forties when the inner self awakens. From then on, the journey is one of becoming the example of balance in appreciation of life.

If a child is an 'Ancient Soul Spirit' then his/her life will be a difficult journey. He/she will embrace all *Soul fears* in a personal quest to find Ascension. During infancy he/she will be extremely psychically sensory, expressing discomfort and joy emotionally, but as he/she grows, they will transform these physical senses into stored information that will be, in turn, correlated with spiritual wisdom that has lain dormant in the deep-subconscious mind. A driving need to find spiritual balance will call for a great deal of understanding and love. In time, this one will become a Sage/Magi of the world. During the early years of their lives, they will be very aware and unusually psychic, but likely to lose this connection in the teenage years. It will return in the 30-40's and become vitally important to them in their quest for spiritual truth. As all the *Soul Fears* entwine, a strong connection with The Creator is forged. God is realized. Thereafter, they will follow many battles between the dark and light in an ever-evolving quest to establish unity that will lead to a world of peace and harmony.

Very rarely, a child in this group may be what is called an *Elder Soul* or *OverSoul*. These children are encoded with all the *Soul Fears* and Most of the *Chief Features*. (See Appendix). Their battles within themselves are to find ways to bring a message of love from God, The Oneness and all ascended beings in ways that will enlighten Mankind. Whatever they do or say will always be benign in appearance, but will have such a profound effect that it will cause a total change in the conscious evolution of individual thinking and subsequent feelings that ensure acknowledgement and attachment to God. Such a being was Mahatma Gandhi. He often stated how hard it was to overcome his negativity in the face of adversities.

Your child can be tested for *Soul Fears* very easily. Simply carry out the following exercises:

- If your child cries when you swing him/her high in the air, then he/she has a *Soul Fear* of *Ascension*
- Should you swing your child down towards the floor, and he/she cries out and clings to you with a look of horror, than your child has a *Soul Fear* of *Descension*. (*Descension is a word I created in 1965 to keep things sounding the same. It seems many now use this word, though it is still not accepted in the dictionary.*)
- If your child cries a lot when you put him/her in the crib, or leave the room for a moment, then your child has a *Soul Fear* of *Separation*.
- During the first six months of life, notice how your child reacts around strangers. If he/she cries when passed to be held by someone other than family, then the *Soul fear* of *Assimilation* is apparent.

- If your child seems to cry a lot when disturbed, such as sleeping, or eating, playing, then your child has a *Soul Fear* of *Unconditional Love*.

While the rest of the *Soul Structure* may be very different in each child, they will seek out those who are 'In the Mirror Image' of their lessons. As we say "Birds of a feather, flock together." Your child will make friends with others who are interested in the same games, talents, schools, etc. During those formative years, he/she will grow to accommodate changes and expect to make changes themselves, but not without a prior battle to hold onto their familiar history. Be patient with him/her as they face their frustrations. In time they will learn to be confident.

Though each child has their own unique programming within their own *Soul Structure*, they do have some facets in common with one another. They are all outwardly dealing with the *Soul Fears of Ascension, Descension* and *Unconditional Love which other people toss at them*. In other words, they are easily influenced by people's emotions and are in a constant battle between acceptance and denial while struggling to find love. Most Star Children have a *Goal of Growth*. (See Appendix).

This *Goal of Growth* is basically a lesson in comprehension and confusion. Because each child is unique, there can be no precise plan to the evolution of Mankind, hence the freedom of spirit to express itself through daily routines etc. Whatever, each Star Child decides to do, you can be sure of one thing, he/she will not settle for just any old thing.

Evolution has come a long way and history has recorded our spiritual evolvement. Today, these children in this group will be the living example of the past reborn so that we may understand the beauty of our lives and the planet we live upon.

Some of these Star Children will re-enact atrocities, while others will lead exemplary lives. In the process of this battle, children from the other groups will play their part, ultimately leading to a better race of Mankind.

As with Star Children born before this time, these children will explore the world in awe and bewilderment, trying desperately to understand their role in it. They will listen and learn quickly and could easily be led into false beliefs and disciplines. By embracing the wrong, they learn about what is right. In their search for the truth, they will destroy and rebuild according to the times.

As with all children, pacts are made in the Spirit World to become families. These children have long ago chosen their families. They have planned to be born in this age, and to accept responsibility for the coding of their DNA strand that carries issues from the past. They will delight in old things, and enjoy delving into historic events. They will enjoy hearing about their great-grand parents and any other historic family stories.

While we live in a technological age, these children will still delight in comparing the past with the present while enjoying the exploration of things to come. They have a love of the land, the people on it and all that nature provides us. They are the preservers during the changes that are to come.

Support Your Star Child Positively By:

- Physically touching your child while explaining what something is.
- Always talking and explaining your feelings and actions.
- Allowing your child to have a point of view.
- Discussing everything as it occurs and ironing out problems.
- Supporting challenges.
- Understanding your child's body language.
- Drawing out any signs of withdrawal through involvement activities.
- Establishing a strong emotional link.
- Accepting that your child will see you in a different light.
- Being organized with routines.
- Encouraging independence with lots of praise.

- Teaching the importance of being true to his/her feelings.
- Encouraging reading and interest in the sciences.
- Always following through with promises.
- Enforcing structures that give good moral support.
- Teaching good ethics with a spiritual base.
- Encouraging debates and discussions.
- Giving chores and responsibilities as lessons.
- Educating about drugs and their misuse.
- Providing authoritative structures with reasonable rules.
- Refusing to accept failure

Protecting Your Star Child From Negativity By:

- Avoiding yelling and screaming at them or at others in front of them.
- Not ignoring your child.
- Refraining from denying his/her reality or lying to them.
- Desisting from ignoring their input in family situations.
- Avoid saying one thing and doing another.
- Keeping from making one plan and then cancelling it without very good reasons.
- Preventing excuses to leave your child with strangers.
- Refusing to ignore his/her need for discussion. Be alert to emotional changes.
- Preventing yourself from creating dominating dramatic scenes that harm.
- Not punishing your child for being different.
- Refraining from accusing them of being something they are not.

- Avoiding assuming you know all about him/her.
- Protecting your child from making bad influential friendships
- Keeping from insisting you know best without explanations.
- Avoiding shutting this child in a room as punishment, or leaving them to wallow there.
- Absolutely refraining from imposing responsibilities that are your own upon them.
- Abstaining from imposing your reality, habits and ways, while overriding theirs.
- Keeping from denying their creative skills and talents.
- Desisting from judging their love of the world and humanity.

Chapter Six

Indigo Child

Children Who Are Natural Mediums

These unique children are born with the vibration of the color 'indigo' a beautiful hue of blue that is a dominant color within the realms of spiritual consciousness. To understand this child, it is necessary to remove Earthly consciousness and judgment.

In the Spirit World there is no judgment. Everything is accepted as relevant or irrelevant to one's point-of-view within a given moment. Each individual event produces a moment of acceptance and revelation. In this way, a spirit learns to trust itself in making choices by paying great attention to their individual sense of feeling. There is no logical or critical attitude at any time in this state of awareness

It is important to understand that the colors of a spirit's aura have an entirely different effect on the colors of the earthly-self, so when a spirit decides to incarnate as an Indigo Child, there is a twist in the way he/she 'sees' things. Their psychic senses are acute, but dominated by Psychometry; the psychic sense of touch and emotional feeling. So, these children feel their way through life. Feeling is sensory, in that they define acceptability as pleasing or discomforting. They first investigate everything they encounter with an expectation of it being pleasurable. Should their experience turn out to be negative, then emotional imbalance and a distortion of reality is likely.

These children are in form right now to mirror-image The Creator. Their personal journey is one of discovery. They seek support and opportunities to create new things that will in some way bring pleasure to themselves and to others. Everything they create is perfect for them in their own way. Should the rest of the world question or doubt them, they will seek a way to share themselves within The Divine Order of God's Creativity. Should they not be accepted, they will most certainly withdraw into themselves and develop traits that are highly likely to be very self-destructive.

As stated earlier, no child is born without an established spiritual bond. But in spite of this, these *Blue/Indigo Children* are born with exceptionally potent emotions and an extreme sense of psychometry (intuitive sense of being able to sense energy transmitted by others that bombards their Aura), which in turn, naturally develops acute emotional needs to belong to society. Their personalities develop as a result of certain aspects encoded into their deep-subconscious. These aspects are a part of the *Soul Structure*. 'Blue' or 'Indigo' babies have one very strong aspect in common: They all have a strong *Mode of Passion*. (See Appendix). This gives each child a great desire to share their emotions in the most joyful way possible. By being stimulated through physical exchanges of energy on Earth, they can use psychometry to activate their inner emotional joy or turmoil in an effort to express unconditional love and acceptance of self and others. Their sense of awareness of photons, 'light' hitting their aura, is extremely acute. The Universe creates light and so do we. (Photons are created in outer space, while we create them from inner space.) These tiny specks of light are created by friction that emanates from the Auric Field of individuals, which can be seen by the naked eye. Indigo Children can see these photons sparkle and interpret them. They either like or hate people, depending on the negative or positive charge generated by that individual.

Individual aspects of their *Soul Structures* are identifiable by the way they function in this world, making them unique; however, they do have several aspects in common with one another. They all work with a subconscious focus on understanding *Rejection* (See Appendix) throughout their entire life. Their final awareness before death is to understand that unconditional love does not come at a price, but rather through acceptance of a child's personality and spiritual presence within; in other words, to live and believe in the goodness of humanity.

Each of these Indigo Children also has the *Attitude* (see Appendix) of the *Spiritualist* encoded within their *Soul Structure*. This *Attitude* provides a direct spiritual connection with God and everyone else as they discover and understand their own emotional and spiritual experiences. From a spiritual point-of-view, these children are able to bring innate wisdom from *The Oneness* (Universal Consciousness where all things are known) into the realm of the conscious mind. Through the development of the other psychic senses, they will activate their *Spiritual Center,* (See Appendix) encoded within their *Soul Structure*. All children in this group have a *Center of Higher Intellect,* (See Appendix) which when used with confidence, may well be expressed though emotional outbursts that can be negative or positive, relevant to the issue that occurs.

Depending upon the child's need to learn, the *Soul Structure* may have a *Chief Feature* (See Appendix) *of Self-destruction* or *Self-deprecation*. If *Self-destruction* is chosen, then it is possible and likely that this child will die young, but with *Self-deprecation* they will certainly live to be very old. However, a spirit is free to choose any of the *Chief Features* to work with. An Indigo Child can indeed 'blue' and hard to live with if negative.

These *Blue/Indigo babies* are spiritually our future. They bring to us the development of creative ideas and discoveries. They show us the importance of being aware of ourselves and the way we express our emotions in every moment of our lives. These individuals become the counselors of the future once they find their own personal balance. They bring messages of enlightenment through the arts, medicine and the media.

In their quest to bring peace to the world, they will explore every avenue available to find explanations. They will simplify and demonstrate those explanations in negative or positive ways. In this way they will unleash an awakening of emotional wisdom upon Mankind.

An Indigo' Child's influence during Pregnancy

In the Spirit World, a child establishes a spiritual bond with their mother prior to birth. As with the 'Hero - extra DNA child' of chapter three, these children do not enter the world lightly. The spirit of each child is so attuned to 'The Oneness,' where all things are known, that they willingly take on all the "Pains of the World." upon themselves Those pains are the emotional fears that Mankind has suffered. By entering the world with difficulty, either in body, mind or emotions, they are able to begin an early awareness of the suffering ways of the world. It is, therefore, not unusual for the mother of the child-to-be to have some kind of problem before or during the pregnancy.

These children awaken in the womb by sharing the mother's total state of negativity. The more negativity there is in the mother, the more likelihood of a life of suffering through negative experiences. Therefore, it is important for the mother of a child in this group to overcome her emotional state and deal with her problems in a practical way that will teach him/her to see that positive solutions are available.

If the mother is in a bad relationship, or uncomfortable lifestyle, she will be emotionally torn about how to deal with her problems. Even the pregnancy could be a problem. She will be mentally driven to solve her problems while emotionally confused about her decisions.

If she seems to have no problems, but has a mind that is over-active, then it will be hard to rest and relax. This causes her child within the womb to develop a need to be nurtured. In the mirror-image of this need, the mother will desire to nurture her unborn child. She will try to plan the perfect birth and perfect life for her child. Since no mother plans a bad life for her child, the consequence of birthing this type of child is to learn how to aspire towards perfection through a mutual need for emotional bonding.

If the mother's lifestyle appears to be perfect, it may well be that deeper problems lie beneath the surface. The pregnancy often brings deep issues to the surface. In this way, the mother's basic instinct for survival will stimulate her child to develop co-dependent traits.

The above information sounds negative, but in reality, this is a very positive way for this type of child to grow and develop leadership in such times of Earthly change. It is the challenge of overcoming negativity that awakens the spirit of the Indigo Child in later years with powerful emotional stimuli, to become a living, walking example of unconditional love which will bring a positive lesson of harmony to the world.

A pregnancy may well be unexpected, or happening in a most unusual way. Such examples are: artificial insemination, rape, middle age and teenage pregnancies, or during a time of break-up in a relationship. Whatever the mother's rationale may be, her heart will be full and open to this union of her spirit with the spirit of the unborn child.

It should be understood and accepted that no pregnancy, even terminated ones, are unexpected. From a spiritual point-of-view, it was planned the way it is. Usually, a spirit that only shares growth and energy in the womb for a short time is working with a *Goal of Rejection* (See Appendix) which is stimulated into action by the other aspects of the *Soul Structure*. Occasionally a Spirit Guide will attach through pregnancy in a *Goal of Acceptance* to help the mother outgrow her negativity. Since they do not intend to be born, they withdraw once the healing work is done and the mother later transforms her energy. This uplifting preparation can then set the stage for an Indigo Child to be born. Some Indigo babies follow-up on this miscarriage as soon as their mother-to-be becomes pregnant again. They make sure they are born with the right beginnings for them to grow through unlearned lessons.

Assuming that her pregnancy is normal and continues, she will be overcome with a P*assion* (See Appendix) to make things emotionally right in her life. She will reject all those things that have upset her. It will be important to gain support from those she loves, and she will demand a great deal of herself in learning to trust her emotions and the subsequent decisions she makes. During that time she may reject some of her ideas and awaken to new realizations about herself.

It will be hard for her to recognize that she is teaching her child to stabilize emotions by intelligently understanding her life which may well cause her to be up and down emotionally. This kind of behavior will stimulate the *Soul Structure* of her child to seek spiritual answers.

At times, she may feel highly motivated, while at other times she may feel miserable, depressed and unattractive. Her yoyo emotions may be completely out of character as she harmonizes her *Soul Structure* with the *Soul Structure* of her baby. Yet, at other times, she may seem to be highly expressive and easily sensitive to those in need around her. Her longing to share and be accepted will be uppermost in her mind.

Indigo babies are encoded with Spiritual Higher Intellectual abilities in order to bring Divine Wisdom into their Earthly life. As a mother accepts her child's *Center of Higher Intellectual Self* (See Appendix) within her own consciousness, she will be motivated to study paranormal or psychological aspects of herself. Those who know her well will consider her going 'out on a limb' in her quest to find herself.

There are exceptions, but most Blue/Indigo babies have chosen one of two *Chief Features* encoded into their *Soul Structure*. However, they can borrow the other Chief Features to develop their character as they mature. For example: they could appear to be highly greedy or arrogant according to the moment. But, if you look behind the charade you will see the true Chief Feature.

Depending upon the choice of *Chief Feature* (See Appendix) within their child's *Soul Structure*, the mother may develop negative traits in preservation of her own need for survival. Examples are:

1. *Self-destruction* can cause feelings of inadequacy in being a mother, with early desires to terminate the pregnancy or give the baby away.

2. *Self-deprecation* can develop an inability to deal with the pregnancy by denying the presence of the child within the womb and subsequently the beauty of herself.

If her own *Chief feature* is one of the above, she will struggle to keep her child, no matter what the circumstances, which will give them both a very hard but deep and meaningful life together as they deal with one another's issues.

In all cases, she may feel that she is in a predicament that seems to have no solutions. But, in spite of this, if there has been a 'promised seal in Heaven,' then the baby will be born and a promise kept. Any struggles that may present themselves will ultimately be solved.

How To Bond With An Indigo Child In The Womb

If you can honestly say that several of the aspects explained before in this chapter apply to you, then you will instinctively know that you have a 'blue/Indigo' baby.

First recognize that you are your child's teacher and, in turn, that your child is teaching you something about yourself. By problem solving, using your creative skills and talents, your child will learn prior to birth to listen to the creative side of its nature. If you are creative, empathetic and sensitive yourself, then your child will learn from you and be born with acute sensitivity, with his/her psychic sense of Psychometry at optimum level.

Express your emotions freely, and acknowledge that everything you experience is also your child's experience. Talk out your issues aloud. Let your child hear your voice and establish a psychic rapport. It is important to remember that this child is an 'Older Soul' who one day will become a leader. In a subliminal sense, you will receive guidance from your child to help you both plan a better future together.

Develop your own psychic senses, and learn to meditate. Become aware of spiritual presences, such as Guardian Angels. Open up your heart to the power of positive thinking and the manifestation of your dreams. Take stimulating classes and develop your talents in the arts and crafts. Write poetry or keep a diary about how you feel during the pregnancy. Also, do physical exercises that are gentle and insightful. Yoga and Tai Chi are excellent ways for your baby to learn about mastery in the body.

If you are positive, your 'Blue/Indigo Spirit' child will give you a deep sense of unconditional love in return for all your sensitivity to his/her needs. You will feel inspired to create and develop your hopes and dreams and live in great expectations of things to come.

The Birth Of Your Indigo Child

'Blue/Indigo' babies are often lazy in the womb. The journey into this world is a tiring one, and it is likely there will be a long labor. During this time, your baby's *Soul Structure* is being activated to survive. Your birthing pangs will be strong, as described by the saying 'He/she suffers the pains of the world.' You should learn to relax and practice deep breathing while emotionally comforting yourself with every contraction.

Once your Indigo baby enters our world, he/she will immediately look to identify you and his/her father's energy by using their psychic sense of Psychometry. If your child is taken from you both before bonding, then your child will feel abandoned and will try to bond with another person. So, during the cutting of your baby's cord, both you and the father should touch your child, allowing both of your energies to blend with those of your child. This will immediately establish a bond, which will never be broken.

Special Notes About Indigo Children

If an Indigo Child is to be rejected, that is to say, separated from the mother by death or given up for adoption, then it is important that the female relative/adoptee hold the baby immediately after its birth. If human contact is not made, and the child does not bond with anyone, then that child will grow up searching throughout its life for a connection that will bring out feelings of unconditional love for self, God and humanity.

If a child is not nurtured during the first five years of life, then he/she will develop an *Obsessive Compulsive Behavior or Hyperactive Misbehavior* pattern within their personality in order to protect their gentle feelings. Over the years, their intellect will mask inner emotions causing their spirit-self to be suppressed. This creates an intellectual cover, hiding their emotions constantly. They will then carry out unnecessary acts that are *Self-destructive* along with emotional outburst, tantrums and abusive physical attacks on self or others.

Their lack of emotional understanding can also cause them to shift into a state of *Self-deprecation*, where they lack the ability to bond effectively in teenager and adult years. If either of these *Chief Features* (See Appendix) is encoded into their *Soul Structure*, the result is always a negative sense of low self-esteem, projected outward in the form of prostitution of the mind, body and spirit. *Self-deprecation* results in negative emotional manipulation or blackmail to self and others creating inner misery. By demonstrating such pain and suffering, it is hoped that those who know them, learn to find positive solutions to their own problems.

Indigo Children can have any of the *Chief Features* encoded within their *Soul Structure* and will show the many faces that mirror image the ways of the world. Their danger is in losing their way and their identity.

If a child is well supported and encouraged over time, he/she will evolve into a 'messiah state' where they will learn to harmonize their angelic ways with the ways of Earth. Whatever their choice of action, they will be multi-talented and able to express themselves easily. They will be accepted by everyone as examples of unconditional love. The spiritual teacher in them will be manifested.

Tender Moments After Birth With your Indigo Child

During the last stages of labor, allow your pains to become one with the experience of birthing your child and share those pains with the child as he/she makes her way through the birth canal. This will make birthing easier for you both.

While your child is having the umbilical cord cut, he/she should be placed on your stomach with the father's hand on your child's body. During this moment, your baby will bond with both of you. This will establish a complete trust in you both as a family unit.

Your child will be hungry for food and longing to be fed often. Natural breast-feeding will comfort your baby, who will quite frequently fall asleep during his/her feed. This can be emotionally draining for you. So, take time for pleasure in this important part of baby's growth. The more patient and loving you are, the better your child will feel about his/her existence on Earth. Find something pleasing to do while your child is nursing.

During bathing or changing clothes, talk to your child to reassure him/her that the change of temperature, and feelings of separation are normal. This will help develop your baby's personality.

Teach your child to accept separation from you as quickly as possible by introducing bottle feeding as soon as is comfortable for you both. At night, place the baby in the crib, and when he/she cries, comfort them, and then place him/her back in the crib. In this way, you show your child independence and encourage the stimulation of *Goal of Rejection* (See Appendix) to establish a strong ability to develop autonomy.

Indigo Babies Have Soul Fears

The choice of your child's *Soul Fears* (See Chapter Three) maybe more than one. Therefore, there can be conflicting signs in the way he/she develops his/her personality and character. Whatever, they are, they will be obvious.

With a *Soul fear of Ascension* your child will try desperately to be 'Godlike' and to be the ombudsman or 'fixer child" in an attempt to find comfort and success. Their emotions will run high and wide.

With a *Soul fear of Descension* your child in this group will be emotionally impassioned to find answers that will prevent failure in both their own eyes and the eyes of others. Opinions of others will affect them to try harder, but no matter how hard they try, they will still seek unconditional love to make them feel safe.

If a child in this group has a *Soul fear of Separation* it will be especially hard for them to stand alone, and make decisions without seeking emotional support. If they are rejected or judged, they will develop a desire to discover other ways to function using the *Spiritualist Attitude* (See Appendix) aspect of the *Soul Structure,* which causes them to investigate ways and means of self-discovery.

For children in this group, the hardest *Soul Fear* to deal with is the *Soul Fear of Divine Love.* Independence does not come easily. They will constantly reject themselves or anyone else who does not support their need to feel safe. They will judge and condemn themselves and others in their quest to understand the freedom of expressing love.

Indigo Children with a *Soul fear of Assimilation* are likely to lose themselves in other people's ideas and plans. Given time, they will become good empathetic and sympathetic supporters to others, but lack their own individual identity while failing to develop their own dreams. They will develop a complex personality made from everyone they meet.

All children in this group desire to experience *Unconditional Love* and will accordingly chose this *Soul Fear* above all others. But, if it stands alone, they are likely to be abused and misused or praised and adored in states of heightened hypnotic awareness. This swinging of the pendulum between the heights of elation and the depths of despair will provide a spiritual lesson. In physical terms, it will be difficult to find an emotional balance.

Because all Indigo Children have a high I.Q., it must be assumed that their experiences are beyond the normal. With several *Soul Fears* entwined, it is possible to make great changes in their perceptions in the way they see their world and the people in it.

The older the Soul of the individual, the more *Soul Fears* will be chosen. This will help them to become truly awakened to both the light and dark aspects of humanity. In this way, a child can suffer and understand a state of *Oneness* and become a 'light to the world' as a living example of joy.

If there is no emotional support, and the child develops a mentality of resistance, then the *Chief Features* of *Self-destruction, Self-deprecation, Stubbornness or Martyrdom* will result. In this way, the child is a living example of pain for others to see and learn from. By using these *Chief Features* a child can move him/herself into awareness of negativity and through identification, can awaken to positive alternatives.

If your child is in this group, you will find a great source of comfort in their trust in you. In times of need, they will find the strength and wisdom to become your parent. So, encourage your child to help themselves first, and then to help others later.

Support Your Indigo Child Positively By:

- Being aware of your child's spirit at all times. You are talking to a wise one.
- Exploring everything emotionally, while expressing mental balance.
- Giving hugs and kisses as often as required.
- Demonstrating what you feel calmly and reassuringly.
- Clarifying "yes" and "no" and stick to it.
- Praising skills and talents often.
- Feeling your child's fears and comfort them.
- Allowing and exploring emotional outburst with insight and inspiration.
- Encouraging joyous excitement and success.
- Helping your child to develop latent talents with passion and enthusiasm.
- Dispelling qualms of imbalance with creative objectivity.

- Encouraging decision making from a sensory point-of-view.
- Discouraging self-destructive actions and verbal dialogue.
- Discouraging comparative/competitive behavior.
- Showing examples of peace and calmness in times of stress.
- Developing an inquisitive nature.
- Showing how investigation results in knowledge and experience.
- Teaching hands-on activities with visual toys along with emotional expression.
- Validating emotions with logical satisfying conclusions.
- Showing how rejection can create protection and safety.
- Giving good schooling in the use of the body.
- Teaching preventative violence and compassion.
- Acknowledging your child's psychic abilities always.
- Taking time to have long discussions when called for.
- Allowing your child to develop naturally at their own pace.
- Placing education emphasis on the arts and sports – intellectual development will follow.

Protect Your Indigo Child From Negativity By:

- Not denying your child's feelings.
- Refusing to accuse or judge them without clear facts.
- Abstaining from speaking or acting for them.
- Not ignoring their needs, even when you are busy.
- Preventing yourself from denying their reality.
- Avoiding pushing them away emotionally at all times.
- Not isolating or punishing them with slaps.
- Desist from avoiding the truth.

- Avoiding telling your child that he/she is incorrect.
- Keeping from disregarding their psychic impressions no matter how weird.
- Being aware of signals of needs for help.
- Avoiding brow beating them with logic only.
- Rejecting support when he/she is whining or crying for attention. Look for answers.
- Not covering up or pretending something is right when it is wrong.
- Refusing to destroy their efforts with negative criticism.
- Avoiding overly fussing about his/her abilities.
- Not placing worries and responsibilities on young shoulders.
- Refusing to drink, gamble or do drugs in their presence.
- Withholding the details of your adult problems, while admitting you do have problems which you intend to solve.
- Keeping from taking over or dominating any of their projects.
- Avoiding arguing in front of this child.
- Refusing to argue with this type of child, or to justify your rules.
- Desist from encouraging your child to be lazy by doing things for them.
- Avoiding telling them that they are too slow or too fast.
- Never feeling inferior to your child's growing inspirational insights, especially in the teenager years.

Chapter Seven

Crystal Child

Children Of The Earth

It may seem strange to think of a child as a piece of crystal. But first let us look at the formation of any crystal from the Earth. It is often multi-faceted, can be opaque or clear, large or small and very complex in its compound makeup. But, whatever its nature, it is unique unto itself. There is never another crystal that is identical, even though it may look the same. Such is the nature of each child in this group. They are all different, yet this uniqueness bonds them together in a firm commitment to establish some form of harmony within our varied personal existences. They are the link that brings all of us together despite our religious, spiritual and cultural differences.

This group of Crystal Children is all encoded with the spiritual *Soul Fears* of *Assimilation and Separation*. Their spirit's sole journey in life is to find an extremely personal way to blend and absorb everything they encounter directly into their own heart. In this way, they are bonded to the ways of the Earth and with the people who live on it. They feel everything and everyone's emotions right across the entire spectrum of light and dark consciousness. They can be likened to a sounding board that reverberates with every touch. When bombarded with 'light' and cosmic photons they quiver and tingle with delight or sorrow. Every day brings them a multitude of experiences that stimulate their *Soul Structure* to resonate with others until they find unity by connecting with every living thing on Earth, the Earth itself, the Constellations and the Spirit Entities in the *The Oneness*. (God's entire creation). The Yang and Yin of these aspects of Mankind is mirror imaged into him/her constantly. Their spiritual quest is to become flexible; shifting in and out of balance until harmony of the mind, body and spirit is found, which inevitably leads to *Assimilation*. (See Appendix).

On a conscious level of awareness, each of these Crystal Children will be born with a deep-seated fear of friendship. They will however, seek out new acquaintances constantly. Their search is always to find someone who is just like them. In the initial discovery of friendship, they will see aspects of themselves in another and bond on those levels, but in time they will begin to see the flaws that lie within their friends and loved ones and then see that flaw mirrored inside him or her self, which will immediately resonate an imbalance within the Crystal Child.

All of these Crystal Children have a different coding within their own *Soul Structure* that gives them a very dominant personality, whether they perform positively or negatively. Their ultimate goal is to get your attention and then to give you a new point of view about yourself and the things you do. They are able to use their choice of three of the Modes: Power, Caution, Passion, Repression, Aggression, Perseverance and Observation (See Appendix) to develop a very strong will, often one that cannot be broken. Their character can be fixed throughout their life, should they have a negative point of view with low self-esteem.

They quickly abhor anyone or anything that is judgmental, yet despite this, they will judge themselves. Their quest to find the perfect home, country, child, animal etc., is always in the front of their mind. They will leave no stone unturned in their overpowering ways to be satisfied. Should they be exposed to the 'wicked' ways of the world, they will want justice and forgiveness for themselves and others.

These children in his group do not seek revenge, but do seek order. They will build and organize metaphysical churches, hospitals, schools and the like. They will embrace all religions, creative talents and skills as well as scientific research that will aid in curing diseases, mental and emotional maladies as well as erasing the use of drugs, alcohol and torture.

Each Crystal Child in this group can be impatient and restless. Their need to get things done will push others or self to the edge. Their tireless quest for obtaining the most comforting of environments will push them into many relationships, though they may not produce large families themselves. Unfortunately their battle with the *Soul Fears* can cause deep depression, anxiety and loss of self-esteem, worth and values. When such negativities occur, they will congregate together with others who are of like mind in hope of finding inner peace.

Every child in this group feels connected to Divine Consciousness and is automatically a channel for Spirit Guides, Angels and God through their connection to their OverSoul (Spiritual Soul Group). Their acceptance of humanity and their caretaker need to help Mankind move along is strong. Their nature to be missionaries in the quest for harmony, education and understanding will far outweigh political views. They will only accept a world where everything can be explored, shared and integrated. This strong connection with The Oneness is stimulated by programming from the *Higher Emotional Center* (See Appendix) and is the channel through which they feel God and all that is created. They are full of visions and often highly psychic.

A Crystal Child's Influence During Pregnancy

I have chosen to write this part as though I am speaking to you directly. This child requires personal attention just as you do. Preparation for this life has long been arranged before your grandmother was born. A spirit has to wait until the moment for its arrival into this world is perfect. Perfect, that is, for all the lessons she/he needs to acquire within the first 15 years of life.

Every friend-to-be, grandparent, uncle and aunt, teacher and so on are all old friends from previous lives who have volunteered to be on Earth in this period of history to help transform this world. This Group of Crystal Children will work with the other four groups who will interact according to their own individual coding to help any Crystal Child to become a true

'Child of God.' That is in the biblical sense! These children accept everyone on face value at first, but later on, deeper levels of awareness release wisdom from their past lives.

You may have been hoping and dreaming of having a child since you yourself were only 5 years old in preparation for the spirit who will be your future Crystal Child. You probably had a fantasy of your 'Prince Charming' riding through the sunset to save you from your miserable life. You ideas of a perfect child have been well defined and you expect to get what you want; a sweet boy or girl. Especially, a boy, who can save you if your 'Beloved' fails you.

During your life, your child-to-be has followed you around, keeping a close eye on everything you do. It may often feel as though you have an Angel watching over you. In truth, this future child of yours is protecting you, ensuring that you will indeed bring this spirit into Earthly life. At the same time, this intended child is preparing through your eyes to know the ways of the people on Earth and the nature of the Earth itself. In other words, they want to fit in right away with what is happening in order to be able to create changes as soon as is humanly possible.

When the pregnancy occurs, you will immediately feel delighted and begins to search for the perfect things to make you and your baby comfortable. Develop crafts and put great pride in making everything yourself. Delight in learning to knit, sew, paint, design and even set up sound effects to make your baby's environment perfect. As the pregnancy progresses, your heart will be full of joy for which you will thank God daily, while always planning in some way to ensure safety for yourself, baby and family.

As your body grows with the baby, so does unconditional love for this child evolve. You may hope for a girl, but may know intuitively that it is a boy or vice versa before proof from medical tests. At the same time, the spirit of your unborn child is busy exercising in preparation for the delivery. In the later months of pregnancy, you will get little sleep as the baby kicks around in her/his internal gym.

Throughout this pregnancy, mother and child are as one. Sometimes it may be hard for you to separate your emotions from those of your baby. It may seem as though you carry your own life in your womb. As your spirit and the spirit of your child bond on a physical level, old past lives will flow into your dreams. Turn to metaphysics to find the meaning of those dreams. Read to research and challenge yourself to find answers; notice how desperately important it is to try and understand the spiritual bond you have with your Crystal Child. Questions will flood your mind as your spirit-self aligns with the spirit of your baby-to-be. You will wonder: "Have I birthed this child before? Was this baby, my father, mother, sister, brother in another life?"

As the pregnancy matures, you will develop an important need to be alone for long periods of time to soul search your private thoughts and innermost emotional deep-self. In this way, you will satisfy needs to a*ssimilate* (see Appendix) with your unborn child, as the baby grows stronger. By the time your child is born you will both have a very special relationship. Should it be impossible to keep the baby, there will be a spiritual bond tying you together all your life. Perhaps, somewhere in the future, there will be a stimulant that may bring you both back together.

How To Bond With Your Crystal Child In The Womb

Your childhood dream has come true and you are pregnant. Wow! Have a party. Now sit down and begin to plan everything in these early days of your first trimester. As you do your preparation work, take time to enjoy rests and to gently rub your abdomen to make physical contact with your child.

Always remember that everything you do is being observed by your child. Listen to your intuition. If you suddenly change your mind and want to put another 'color' into your work, recognize your child is helping you choose.

At night, rub your abdomen with olive oil and ease yourself into a comfortable bath/chair and relax after a hard day's work. In this way, you will teach your child that rejuvenation is the thing to do.

At work or if active in the home, take time to organize. The more you get done in an effective way, the quicker your child will learn from you. His/her spirit likes things to be in place.

Take gentle body/foot massage therapies or yoga classes until the third trimester is completed. This will keep your energies flowing and your emotions on a high. Then spend the remainder of your pregnancy planning the first year of your baby's future. Check out kindergarten, schools, daycare, nanny's etc. You want to be sure that your child knows there is a plan.

Have good discussions with your mate to ensure his understanding of your need to have quiet time. When baby arrives, there will be a lot to do and he may feel neglected. So, to avoid this, make sure you know what he needs, too.

During the entire pregnancy, have your husband stroke your stomach and speak to the child often so that his voice becomes familiar. This can also be done by any siblings.

This new child will require a good spiritual/religious training, so check out the local churches and clinics where your ideas and views are considered and followed. It will be important for the spirit of your child to have a conscious connection to *The Oneness*.

The Birth Of Your Crystal Child

Approximately one week before you go into labor, you will sense a need to prepare. Your baby's spirit is warning you he/she is ready. You may get false alarm pains. Don't panic! Your baby's *Soul Structure* is being activated in preparation for his/her psychic senses to kick in at the moment of birth.

Despite all this preparation for the birth, this child could arrive early or late depending on their own Spiritual *Goal*. The following is something you will notice at the time of birth that will let you know their *Goal*:

> *Acceptance*: Born with eyes open wide and a deep sense of peace.
>
> *Rejection*: Born with eyes open, but crying or making sucking noises
>
> *Growth*: Born with definite nervous shakes, trembles and crying
>
> *Retardation*: Born with some level of deformity or imperfection
>
> *Dominance*: Born with complications (actions that demand attention)
>
> *Submission*: Born sleepy and non-responsive to touch
>
> *Stagnation*: Born early with some aspects of self under-developed.

Since you will not be aware of what your child's *Goal* is, it is important to prepare for anything. This is a part of the evolution of your bond with your child. To accept everything that they are and to do the best that you can is to acknowledge your child's connection with his/her OverSoul's *Goal* of *Acceptance*.

So, during labor, breathe deeply often. Try to use yogic meditation to transfer your mind from the birth pains into another part of your body. Relax every muscle throughout after each contraction.

During the second part of labor, speak to your child in your mind. Encourage them to help you push them out. Encourage them to push themselves too. With each strong contraction, say aloud, "We push together." Though others around you may not understand what is happening, it is only important that you do. This combined effort in the birthing process will activate your child's *Soul Structure* in a very positive way.

As soon as the baby is born, hold and kiss his/hers face and hands. These first touches connect his/her brain to their own sense of human touch and to your support. Immediately after that, your child's father should take your baby and do the same thing. After that the medical nursing staff can do their work in cleaning you and baby.

Special Notes About Crystal Children

It is not unusual for a Crystal Child to be born into circumstances where the mother is very young, requiring the child to be given up for adoption. Or, perhaps the child is unable to spend time with their mother as she is always working to make ends meet. Even more common, is the presence of older siblings who take attention away from this Crystal Child. Whatever the circumstance, this baby is a survivor and will find a way to continue to fight for his/her own space, and through that struggle, to learn *Separation* and later *Assimilation* leading to *Acceptance* of *Divine Love* (See Appendix).

Since the ultimate lesson of love is the root of Crystal Children's existence, their mothers will start off their pregnancy with great hope and joy. By the time she has reached full term and is ready to deliver, she will be more distant and moody. After the birth, she may fall into a state of rejection or depression, having survived the pangs of labor. These emotional imbalances are a direct reflection of the *Soul Fear* of *Separation*. Her spirit has separated from the spirit of her child. The unity of two beings, having been one, has become two individuals with different *Soul Structures* who are now on two separate pathways. Both mother and baby will crave that unity again.

Over the years that follow they will look to one another for support and understanding. This separation may or may not be obvious, depending on the circumstances of their lives. Each feels as though there is a hole in their heart that needs to be filled. Through the natural separation from one another at the time of birth, encoded lessons are stimulated from the deep-subconscious into conscious creating a desire to be loved.

Both will develop a tendency to be co-dependent and will spend the greater part of their lives seeking that union with someone else other than their mother or child. As the years unfold, these Crystal Children will constructively create a new world in which their children will grow up, content in knowing that Universal Love is unconditional and that perhaps a perfect world is very close!

Tender Moments after Birth With Your Crystal Child

Since your baby has been checked, wiped down, and is wrapped in clean linen, he/ she should be placed on your breast to suckle, even if nothing is yet flowing. This first act of being fed is an important stimulus from the mouth to the brain. It will activate the deep-subconscious to release in-built coding for survival and lessons acquired from past lives.

As soon as you are able, it is important that you make contact with your child's base (Root Chakra) by being the one to cleanse their bottom when he/she has their first evacuation. This will activate the Root to Crown Chakra, thus ensuring your baby's full acceptance of their existing life in this present time. You don't want past lives to be activated at this time. Motherly contact is very important during the first 24 hrs of life. If the father is the only one present, then this will do, but it would be better still for the maternal grandmother to give love and touching because she has similar *Soul Structure* coding as the child's mother.

Everything that your child needs should be done by you as much as possible. Breast feeding is essential for as long as possible. If you have some post-natal problems, or emotional mental issues, then allow your mother to help. Bottle feeding is alright, though there will probably be some weaning issues, along with bodily changes which may include abreactions to new milk.

Mother-in-law can be a second choice as nurturer. However, her coding in her *Soul Structure* may be greatly different from yours. Though she may take care of baby in similar ways as you would, your child will feel her energy as foreign and will immediately try to bond with her. Together they will form a collective energy that will be different from your own. If your child is in her care for very long, she will influence the baby to follow her ways. Be wise in your choice of care-givers. Overcome any depression you may have and take charge of your baby as soon as you can.

Your Crystal baby will be extremely sensitive to your moods. Try to be happy and positive, even though you may have post-natal stress. As stated earlier, Crystal Children are highly sensitive to the electro-magnetic pull of the Earth and to the energies of people and animals, so they will be busy in their first hour of life attuning to the people, smells and energy of everyone in the room; even the whole building. When he/she is older, they will investigate everything they see and mirror image the emotions of those they meet. So, if you are in a bad mood, so will your baby be. Be happy and try to see the bright things in everything that happens.

Both Mother and Grandmother's support during the following three months would be best for your child if you are unable to cope. Your baby's need for fatherly support will develop as soon as awareness of self-will occurs somewhere around 18 months old. Prior to that, the first 18 months of your Crystal Child's life will be focused on female bonding. However, it is important that the male influence be present as a support factor. You will need your husband to nurture and give physical support to both of you in many ways. As new parents, you should both develop an encompassing relationship that nurtures your child for a few hours a day, though not constantly. Remember, the *Soul Fears* of *Separation* and *Assimilation* are both working in a tug of war. By spending short times with the two/three of you together, your child learns trust. When left alone for short periods without either of you in close contact, he/she will learn independence.

Your Crystal Child will want a routine, so feed, bathe and put the child to sleep by the clock and not on demand. In no time at all, he/she will learn to accept these routines and be upset if they are broken. If you are traveling, try to work with these routines as best as you can.

It should be remembered that a Crystal Child has come to Earth to learn to integrate in unconditional love, thus facing and embracing the *Soul Fear* of *Divine Love* but this takes time because of the primary *Soul Fears* that have been encoded. So, take simple steps to show them how to become independent; to trust their choices of what they do and who they do it with.

This very psychic child will demand a great many explanations about life in general and you will need to have most of those answers!

Crystal Babies Have Soul Fears

Crystal Children have the *Soul Fear* of *Separation* in common and when it kicks in they will feel isolated and disconnected and after much suffering, will move on in their quest to find love from someone else.

Every Crystal Child will feel as though they are starving for love and not recognized for who they are. They will leave no stone unturned in their quest to understand why they feel the way they do. Their psyche will develop strong counseling and healing skills at a very young age. The more they feel isolated, the stronger they become. Ultimately, they master the *Soul Fear* of *Separation*.

In this moment of understanding how isolation has brought them to face their *Fear* of *Separation,* (see Chapter Three) an awakening occurs on a spiritual level that activates Divine Consciousness to use the *Soul Fear* of *Assimilation* (see Chapter Three) in an effort to connect to God. Old needs fade, as new wants transform passion into the power of manifestation. This newfound passion evolves with the adult to want to experience total joy and complete unity through emotional absorption into 'Oneness consciousness' or Divine Wisdom. Only then, can the real purpose of the Crystal Child be revealed. Together, this group will show the world how to face the ultimate *Soul Fear* of *Divine Love*. To be so completely and utterly in acceptance of all things in all forms without judgment creates vulnerability which can result in total panic and withdrawal. However, if that fear is faced, then bliss results. It takes many incarnations to achieve this.

Until that time in the far distant future, each child will do battle between their thoughts and emotions until physical exhaustion causes an emotional state of spiritual surrender. This new inner state of joy then motivates them to become a teacher to the world. They will seek the miserable, lonely, downtrodden and the like to rescue them from their pain. In an effort to save those they find in need, they find themselves, make new laws and instill news ways to perceive life. At this particular time of writing, there are many Crystal Children fighting for this cause.

The *Soul Fear* of *Divine Love* is in itself and of itself. All children from all groups must face their fears of surrender to self, family, friends and strangers, Spirit Guides, Angels and God, not necessarily in that order. Pure love is without judgment. A mother loves her child unconditionally. These Crystal Children are to become living examples of their collective OverSoul. (Other Spiritual fragments/group of entities compiles an OverSoul when united as one).

A collective OverSoul consists of many fragments that have bonded and melded into Oneness and which in turn has its own *Soul Structure*. This is still a point of Ascension away from the ways of Earth. This spiritual coding is embedded within each child. For Crystal Children, their OverSoul's group has a *Goal* of *Acceptance* (See Appendix) at this time. While in embodiment, each child has their own individual Goal encoded within their own *Soul Structure w*hich is subtly affected by the collective conscious of the OverSoul. This could or will affect their conscious decisions in how they mature and develop their work. There is no limit to their skills and talents. If they want something, then they must work hard to get it and perfect how they use it. It is important to them that they are appreciated and that they appreciate themselves.

Notes on the Crystal Childs First Three Years

A great deal was mentioned above during the birthing process and immediate care afterward. I have therefore, decided to expand this section to include care for the first three years of life.

Crystal Children are extremely sensitive to their environment and to the people they live with. If anyone in your environment has problems, your child will respond with an action that will set that mood into play. He/she will manipulate the moment to suit his/her own needs. It is important to establish behavior patterns for your baby to follow.

Once a routine is established, be sure to teach him/her to follow directions when learning something new. Their attention span may be short, but their ability to learn is fast. If he/she gets what you teach in seconds, they may rebel immediately to test you. All Crystal Children are spiritually encoded with a need to master *Separation* in order to break away from earthly rules, but of course, they will come back for more lessons later. This on/off nature can be annoying. They can be stubborn and fixed in their ways. Physically, they are learning to be leaders, which in the mirror image of God, awakens their spiritual independence.

Teach your child to explore new avenues of expression to overcome boredom and to follow discipline. The direct conflict between being independent, yet needing a routine is and will be the bane of his/her life until they awaken to the use and ways of sharing unconditional love. Their awareness of Divine Love is locked into their sense of physical and emotional feelings of contentment.

Your child may be bright and creative, but slow with academics or vice versa. They will be frustrated when they fail. They will torment others who succeed. In playschool they must learn to integrate and socialize. These formative three years are vital in your child's growth. Let your baby become who he/she wants to be, but maintain parental control.

Support Your Crystal Child Positively By:

- Letting them crawl, touch and play with anything that will expand their five senses.
- Gently rubbing their back and abdomen to remove gas.
- Allowing them to play with toys that can be handled easily to avoid temper tantrums.

- Encouraging tactile play with physical contact to learn different sensations i.e. hot and cold, rough and smooth within weeks of birth.
- Giving plenty of fluids when your child is crying.
- Physically touching your child while explaining what something is.
- Using simple expression in explaining your feelings and actions.
- Allowing your child to physically show you what they want.
- Discussing everything and always iron out problems as they occur.
- Supporting challenges with warm sounds in your voice.
- Understanding your child's body language.
- Drawing out frustrations with playful sounds and words i.e. "Oh dear! A booboo!"
- Establishing a strong emotional link with laughter and tickles.
- Accepting that your child watches you and sees you from their point of view. (Watch yourself and your expressions in the mirror).
- Being organized with routines.
- Encouraging independence with lots of praise for tidiness and cleanliness.
- Teaching the importance of being true to his/her feelings.
- Encouraging reading and interest in the arts and sports
- Always following through with promises and changes with joy.
- Enforcing structures that give good moral and ethical behavior support.
- Teaching good religious ethics with a spiritual base in unity.
- Encouraging debates and discussions to show that life has variety.
- Giving chores and responsibilities as lessons in discipline.

- Educating them about drugs and their misuse.
- Providing authoritative structures with reasonable rules.
- Accepting their psychic skills and the things they tell you.
- Investigating their deep feelings and bring them out into the open.
- Always being their friend.
- Loving them no matter what they do.

Protect Your Crystal Child From Negativity By:

- Refrain from shouting, screaming, slamming doors or exposing them to loud music or noises.
- Not forcing food down your baby's mouth.
- Avoiding leaving your child in a dirty diaper.
- Not leaving them alone near water or fire – they will explore it.
- Abstaining from spanking your child.
- Refusing to ignore your child's conversations during any event.
- Not denying his/her reality.
- Refusing to allow anyone to ignore their input in family situations.
- Avoiding saying one thing and do another.
- Keeping your plan firm and not cancelling it without a very good reason.
- Avoiding leaving your child with educators/sitters until they are over one year.
- Desisting from treating them differently from other siblings.
- Avoiding dominating this child with dramatic scenes.
- Preventing yourself from punishing them for doing something differently.

- Desisting from accusing them of being something they are not.
- Not assuming you know all about him/her.
- Keeping away from preventing your child making friends.
- Avoiding insisting you know best without explanations.
- Abstaining from punishing your child with separation from family and friends.
- Not imposing responsibilities that are your own.
- Avoiding imposing your reality upon theirs.
- Not denying their creative skills and talents.
- Avoiding judging their love of the world and humanity.
- Not demonstrate alcohol or substance abuse in front of them.
- Desisting from threatening to throw them out of your life or home.
- Never devaluing what they give you.

Chapter Eight

Liquid Crystal Child

Highly Emotional Child Of The Future

Master Chang, my Master Healer Spirit Guide has said that these children are born with an extremely innate awareness that is always brought directly into the conscious mind as a result of outside stimuli. Their brain patterns are different. Their perception of 'light' and subsequent photon sensitivity (Elementary unit of light created by electromagnetic fields within the aura and in the cosmos) is outstandingly stronger than any of the other groups. They are able to use their highly developed psychic skills to attune to their liquid crystalline coding in their DNA and to their spirit self encoded into the *Soul Structure*. This allows them to know feel and understand every living creature on any level of awareness as well as having the ability to attune to the many features of the Earth.

This aspect of the Liquid Crystal Child's existence allows them to know anyone's objective, points of view etc. By being involved in any situation they will seek to understand and immediately know the answers while others are still searching. These children are natural psychics, with a dominant sense of clairvoyance. They personify the expression "A picture is worth a thousand words!" They are the computer generation. Their brains are advanced computers! Their DNA is dimensionally different. A normal brain is capable of only processing the senses according to conscious and subconscious memories. A Liquid Crystal Child can process their spirit memories from past lives along with the evolution of Mankind and can bring forgotten information into the forefront of their minds. They are the next step in the evolution of the human brain. Energy passes through neurons faster than the speed of light and these Crystal Children are capable of seeing themselves doing this. When they gather together, their telepathic abilities are striking!

At this time of writing, I have not heard anyone else speak about the Liquid Crystal Child. Though seeming to be entirely different from the other four groups of children, they are in fact both the newest and the oldest of Spirits. In this Aquarian Age, they are the new models of childhood, but in Ages gone by, they are the oldest of spirits who have caused effective changes in the world. Until now they have not been given credit for their existence and their work. In bygone days, they were the genius and the madman; the saint and the devil. They incarnated in ages past to assist the other four groups of the time to evolve by simply being fully present and active in the moment. In our time, they are born with a deep-seated sense of Ascension, unlike any other time before where they were simply leaders of conscious awareness, the philosophers of their day.

These children can be conceived to be scary. It is not easy to be with someone who seems to always be right, especially if they are only young children. Their wisdom is not of this world. Their insights come from their spiritual wisdom, which has been coded into their *Soul Structure*.

They will often appear to have conversations with 'no-one,' and insist that they see and hear someone or something that nobody else sees. At the same time, they may utter conversations about the past or future. Advice flows from them like water running down hill. Nothing they say is unacceptable to them. Everything is their reality and they expect everyone to understand them.

Their psychic senses are so acute that they can be very telepathic and empathetic. They may often tell you what you are thinking and feeling in such a way, that it can be uncanny. Or, perhaps they may suddenly tell you about your Spirit Guides and dead relatives without you even saying or thinking anything even remotely relative.

Their insight into other dimensions, the Universal and the Spirit World will be inexplicable, but to them it is real and a very necessary part of their consciousness. All things are possible in their world. They will even talk of alien life forms with authority.

Like any other human on this planet, Liquid Crystal Children have a *Soul Structure*. But theirs is different from most, because they have coding which ensures that their own spirit is uppermost in their consciousness.

In order to learn quickly, these children have a very powerful *Goal of Acceptance* (See Appendix) and are, therefore, able to see the good in even the worst scenario. They may agree in unusual ways, and even condone wrong deeds in an understanding of the changes that are occurring in everyone involved.

These children usually have these aspects of the *Soul Structure* in common: a *Mode of Observation* backed up by the *Attitude of the Realist*. (See Appendix) This causes the child to stop and watch and learn. They have high levels of aptitude and expression while developing an understanding for the need of a balanced point-of-view on life. They also have a coding from the essence of their spiritual being, known as a *Spiritual Center*. In these children, the *Spiritual Center* stores the 'All Knowing' aspects of the *Spiritual Intellectual Higher-Self* history, which manifests in the body to provide a deep innate sense of emotional intuition within the brain. If this kind of intuition is not explained or developed, then mental illness or deformity can result. However, in most cases, in order to develop a balanced nature, this aspect of the *Soul Structure* is also accompanied by a *Spiritual Emotional Lower-Self Center* that creates a grounded rational intellect. What they feel is deep and easily based on reason. (See Appendix).

By having the normal spiritual ascended wisdom encoded into the emotions, the Earthly-self learns through daily intellectual exercises to understand the ways of the world. In this way, the journey of the spirit of a Liquid Crystal Child is tested while on Earth. This test will ultimately lead them to ascend into a new understanding of God's Eternal Love.

Each of these children also has the *Chief Feature of Arrogance* (See Appendix) which will lead them into states of pride and vanity where only their opinions count. However, when mixed with the other aspects of the *Soul Structure*, this *Chief Feature* can lead them to greatness. The remaining aspects of their *Soul Structure* coding will create their individual personality and character.

These highly developed psychic children are an anomaly unto themselves. They often show examples of the unexpected and are not quite sure why they do the things they do. They are the seekers of truth and light and will become the living example of how to focus on true awareness to develop the inner self. They are the voice of *The Oneness* in form and the true connection to the Spirit of God and the manifestation of every reality known or unknown.

In watching and copying, they learn about themselves and others. In time, they become examples of both negative and positive humanity in action. By their arrogant nature, they are able to focus on their way and ignore individual aspects of Mankind that might interfere with their spiritual growth and their sense of destiny.

A Liquid Crystal Child's Influence During Pregnancy

Unlike the other children, their bond with their own mother is a more impersonal connection. While they love dearly, they also long to love others. So, during the pregnancy, they attach to the spiritual lessons of their mother, rather than her more Earthly lessons. This indifference to the Earthly ways often makes it possible for a mother to conceive a child that is not expected or planned either from an emotional point-of-view or a practical one. When the spirit of her unborn child arrives, it will check out the womb, the state of the mother and immediately rearrange everything to suit its journey into this world. The mother's response to this is to rearrange her lifestyle to suit the type of child she will have. This in itself is an *arrogant* state of being that is reflected from her child.

Her body may go through radical bio-chemical, and metabolic changes. She may desire unusual foods that are strangely abnormal from her usual diet. She may become interested in alternative medicine, alternative health, and natural and unusual ways of delivery, such as water birth. Her cycle of life changes as she embraces *The Realist Attitude* in action following the influence from her baby, who is using the *Spiritual Higher Intellectual and Higher Emotional* points of view to influence her about her moment of delivery. (See Appendix)

Liquid Crystal Children are completely aware in the womb and often leave by having out-of-body experience before birth. The mother may feel at times that her child is dormant or not present, and she could go through pangs of fear for her child's safety and arrival. This stimulates her to *observe* her pregnancy with caution. By working with her chosen code of the *Mode of Observation* (See Appendix) she is able to prepare her life for the arrival of her baby in the most appropriate circumstances. She becomes active, adopting aspects of her child's *Soul Coding* on a temporary basis as unity builds between the two.

In her desire to protect her child, she will become strong and antagonistic to anyone who might threaten her pregnancy, whether it be emotionally, mentally or physically. Even the father is likely to be verbally attacked if he steps out of line from the mother's point-of-view. The child's *Soul Structure* is deeply entwined within the mother's *Soul Structure*. They are bonded on levels that normal logic cannot explain.

As she accepts this blend of *Soul Structures*, she changes her character. Those who know her will note and remark about her changes. She will rationalize and explain them away in a material sense, but deep inside, she knows she has changed for the better. Her negativity will disappear. She will see things in a new light, in *acceptance* of her new role as a mother-to-be.

If the mother of this Liquid Crystal Child has a *Soul Structure* that is mostly expressed in its negative aspect, then her changes will be extremely obvious. She will use her adopted *arrogant* nature and her need to overcome her fear, pain, anger and guilt will be paramount. She will try to change everyone around her and may even divorce or move away from those she feels are not a good influence. Her ideas will be sharp and focused as she makes up her mind. She will not look back and may be perceived as cruel and unfeeling.

If she is a mother who is well balanced and able to accept a challenge, she is likely to be older in years and wiser from prior births. She will have a highly developed psyche herself and will be extremely adaptable in re-arranging her affairs to suit herself and her unborn child without destroying anything. She will feel that she is ready for change and will make certain to manifest those changes as soon as is humanly possible.

How To Bond With Your Liquid Crystal Child in Pregnancy

These babies announce their arrival prior to pregnancy and during the first hours of ovulation. If you find yourself awakening in the middle of the night; feel a spirit presence and a need to have a child; then know that you have a very special child coming in who is a very 'Old Soul'. Or, if you have just had a wonderful night with your loved one, and awaken in the morning with a sense of being pregnant, then you better know that you are and that you have a special psychic child growing within your womb.

As the pregnancy advances and you find yourself showing many of the traits explained earlier in this chapter, it will be clear to you that you need to spiritually understand this child.

Meditate each morning and reach into the spiritual mind of your child. This connection is spirit to spirit and not body to body. You may discover past-life experiences, or see images of things to come. Enjoy the show! Everything you see will be symbolic of the lessons you both need to learn.

Acknowledge that your spirits are entwined and that you can learn from one another. Develop latent talents and skills that seem to come from within your heart. Learn to listen to your heart and not your head. This child is teaching you.

Accept that the spirit of your unborn child already psychically knows every second of your entire life. This child exchanges images with you and you must learn to understand those images. In this way, he/she will help you to see and understand your mistakes and successes.

While attuning to your child, you will also become conscious of your own Spirit Guides and the Spirit Guides of your child. There is more going on than you realize. You are both the center of attention. Through you both, there is a Universal shift in consciousness as your *Soul Structures* bond. As your Spirits bond, so your Spirit Guides bond, as do other ascended beings and so on into *The Oneness*.

The only rule here is to expect 'The Unexpected' and to go with the flow of the pregnancy. Resistance is futile! So, enjoy unfolding your personality and embracing the existence of self-awareness in the paranormal.

The Birth Of Your Liquid Crystal Child

Your dates for delivery may be firm, but that does not mean you can plan. Anything can happen, from a premature delivery to a very late one. Or, it could be that circumstances override the normal, such as a sudden need for a cesarean birth. Even still more annoying could be the upset of your routines or the changes in attending physicians. Whatever has occurred, remember to go with the flow. Accept the circumstances and learn from the changes that are happening. It is most important to not worry or tense yourself during the delivery, no matter what is happening. If acceptance is ignored, then the lesson of severe pain will follow.

These children often take a long time to come into the world. Their struggle to live and survive is often long and hard. They begin as they intend to live. They will do things the hard way. So, try to make life easy for yourself by letting yourself relax with the birth as much as is humanly possible. Take frequent deep breaths, pant etc. Do everything according to your pre-natal training and all should go well.

Special Notes About Liquid Crystal Children

If this kind of child is born into a negative environment, then he/she will seek out negative situations throughout life. They will either create them if there are none, or will enter into one to challenge everyone involved. They can be argumentative and temperamental. So, the early days of life are important in formulating habits and rules. These infants should be allowed to cry for periods of time to develop their lungs. Liquid Crystal Children will often over-react to everything and cry often. Though it may be annoying and sap parent's sleep, it will teach a child to be independent and to accept vulnerability when left alone from time to time to just cry. There may be many times that this type of child will be feeling uncomfortable and insecure for no apparent reason. They begin to search for safety as soon as they leave the womb. This innate insecurity will stimulate them to establish their own parameters for behavioral patterns that will develop a highly sensitive personality.

Spiritually, this Liquid Crystal Child is identifying with the World, and is bombarded with the energy that emanates from everyone on all levels. As they become aware of life in all its forms, they will recall ancient times, the good and bad, and subsequently will demonstrate talents acquired from past lives without having had lessons. They are natural philosophers, and can freely give information that will amaze those who listen.

Tender Moments after birth

Once your baby is born, he/she may begin to make sucking noises, or crying extremely loudly. Your child will feel naked and vulnerable immediately. As soon as possible, even before the cord is cut, you should hold him/her in your arms and snuggle closely. This Liquid Crystal Child will be sensitive to everyone present in the room, and will immediately react negatively to anyone who is emotionally unbalanced. So, try to have doctors, nurses, other hospital staff and family members helping you who are emotionally stable. These first few moments of birth will stimulate your child's *Soul Structure* to kick into action. You want your child to make a positive start on life.

In the ensuing days, your new baby may well cry a lot to gain attention. Be sure to stay calm and deal with the small issues that arise, such as lack of breast milk, wrong formula, gas, vomiting etc. Your Liquid Crystal Child's spirit will not take kindly to being in the flesh and will have problems adapting to their body. In the process of acceptance, you will be tested.

If you have fed your child, and he/she is still crying, check for gas and do your best to release it. If there are no medical reasons for the crying, then trust that you are both being tested. Be patient, talk lovingly at all times, and don't over-do the mood swings! If you do, then your psychic-child will mirror image those moods. He/she is learning about emotions.

Because a child in this group is highly psychically developed, it is important to acknowledge that your child feels all your fear, pain, anger and guilt, along with your joy, pleasures and talents. The sooner you start sharing yourself in physical and emotional activities the better. If you provide a harmonious environment for your baby, he/she will lie peacefully in his/her crib once separation is accepted.

Liquid Crystal Children Have Soul Fears

While there is no individual Soul Coding that creates a common link with other Liquid Crystal Children or a common *Soul Fear* that ties them together, they all are most likely to have at least three of the *Soul fears* encoded into their *Soul Structure*. By being heavily burdened with Soul Fears, they are able to identify themselves with others in spiritual ways.

If the *Soul Fears* are: *Ascension, Descension* and *Separation* then this child will struggle between needs to be really successful, while longing to procrastinate and fail. In this way, a fear of being an outcast will test judgment and independence. This type of child can create situations to cause issues of abandonment.

If the *Soul Fears* are: *Ascension, Descension* and *Assimilation* then this child will feel embarrassed when paid too much attention. They will long to be successful, but in achieving their goals will deny their claim to fame. They would rather hide in the crowd and be a part of the masses even though their heart cries out to lead. This is not an easy test to overcome and it can create mental disorders such as Bi-Polar problems. These children feel everyone and cannot separate themselves from the crowd.

If a child has chosen to work with the *Soul Fears* of *Unconditional Love, Assimilation* and *Ascension,* then this child will want to help everyone, by feeling and taking on their pains and troubles as their own, so becoming a martyr. They will see the bad in everything, but want to change it all for the better. They will find themselves in a never-ending stream of responsibilities, until they learn that they can only help themselves to evolve. Only then, can they become a living example of success.

Some children in this group may choose to focus on the *Soul Fears* of *Descension, Assimilation* and *Unconditional love*. If this is the case, they will be psychically pulled to run with outsiders, those who don't belong. Together as a body, these individuals will break the rules and fight the systems that bind them. They will form strong bonds with one another and try desperately to share themselves, their ideas, hopes and dreams with others. Throughout their lives they may feel unloved, misunderstood, and forgotten. But, they will leave a mark on Mankind in some way or another.

These children often choose to mix the *Soul Fear* of *Separation* with any or all the other *Soul Fears,* but probably the most frequently chosen is to mix it with *Unconditional love* and *Ascension*. In this way, the spirit of a child will seek God through the development and use of the five psychic senses. These already highly developed senses take on a new meaning when used diligently in life. The healer manifests along with Christ consciousness. Miracles do happen! But, such wonderful abilities come at a cost. Followers and the like judge, condemn, and victimize these wonderful individuals who give freely. So, throughout their lives, they feel lonely and different from others. In their own quest for spiritual ascension, they find God within themselves and those they meet.

Perhaps a child has four *Soul Fears* encoded. If this is the case, there can be confusion for the observer. They could be angelic one moment, and devil-like the next. Their character may change from moment to moment, or they may seem to be wise and suddenly an idiot. As these *Soul Fears* interact, there is a constant battle roaring within the child. Trying desperately to be special and successful, while failing and falling apart is not uncommon. Confrontation and stubbornness may be obvious, but throughout all these negative swings, there will be wonderful moments of great joy as their spirit shines though with understanding. These 'Old Spirit Souls' pass through stages of evolution in the years that they live; growing from seemingly stupid states of childhood to elderly states of wisdom. They become the teachers of this world.

When the spirit of a child is 'Ancient,' all five Soul Fears will be encoded into this child's *Soul Structure*. He/she will be extremely sensitive to any given situation. Their uncanny ability to heal the sick, counsel and ease the mentally disturbed and spiritually lift the lost and lonely is extraordinary. Their pure presence on this planet is often abused and misused, but despite all this negativity, they ascend in consciousness to bring enlightenment to those who seek their counsel. They may be called 'special' or 'gifted', but ultimately, they are judged by Mankind as too perfect to be accepted. In this way, they leave a lesson for Mankind.

Whatever the coding of these very psychic children, you can be sure that they are creative and talented. Whatever their choice of expression, it will be important for them to receive an accolade from time to time. So, give plenty of praise when it is due. With a wonderful support system, and spiritual guidance, these children will mature into amazing adults who will change the world and all who live on it.

However, being so psychic can come at a price. They may have an overactive mind along with too much emotional stimulation. They may develop a variety of dyslexic traits, autism, along with fears and phobias. In extreme cases they can become mentally depressed and possessed by negative entities. Don't expect the normal with this child. Look for unusual ways to help them. Alternative methods of healing and cleansing are the norm for this child.

Support Your Liquid Crystal Child Positively by:

- Being aware at all times of your child's psychic senses and encouraging them to use these skills.
- Teaching them psychic protection and values.

- Always talking about anything and everything they ask about.
- Exploring all their talents and encouraging them to do more.
- Educating your child to sing, dance and be artistic etc.
- Explaining what is happening physically and emotionally.
- Explaining what is happening spiritually and mentally.
- Listening to their dreams and helping with insight.
- Acknowledging their visions and keeping notes.
- Encouraging their changes in focus and helping redirect focus when necessary.
- Always trying to understand their different point-of-view.
- Being flexible with routines, but disciplined with education.
- Always finding solutions to problems.
- Giving constant encouragement.
- Giving guidance when asked when they ask for it.
- Helping develop a good attitude about mind, body and spirit.
- Always referring to spiritual guidance and Guardian Angels.
- Providing good religious background, with freedom of choice.
- Being a good role model at all times.
- Being there for your child when you are needed most.
- Teaching the importance of emotional expression and a balanced mental attitude.
- Nurturing them always, even unto death.

Protect Your Liquid Crystal Child From Negativity by:

- Not over-reacting or demonstrating panic.
- Avoid denying your child's psychic reality.

- Preventing any situation from downgrade or humiliate your child in public.
- Never blocking their creativity.
- Never shouting or punishing when you are in an aggressive mood.
- Refusing to leave your child with a complete stranger.
- Making sure to not ignore their guidance, however trivial.
- Refraining from slapping this child as it amounts to physical abuse.
- Refusing to make a promise that you may break.
- Avoiding your own denial about your own reality.
- Preventing anyone from shutting your child away when visitors call.
- Never leaving your child in a quandary.
- Assuming understanding or ignoring teaching about the dangers of the world.
- Not ignoring their fears.
- Refusing to lie or pretend to them.
- Not judging them incapable.
- Facing negative psychic behavior – seek guidance.

PART TWO

BRINGING UP BABY

Chapter Nine

Which Group Do You Belong To?

Since there have been very many periods in all of history with the five groups of children appearing to transform Mankind, we need now to only focus on the last century. However, it should be noted that there have always been five types of children with a Spiritual Lower-Self focus on Descension. Through 'falling from grace' there was always an evolvement towards Ascension. Recorded history has shown how Mankind has been destructive and unappreciative of their abilities and the world they lived in. They believed in separation and in dominance. Judgment was rampant everywhere. God was believed to be a wrathful God. Now for the first time, all five groups have consciously accepted work on improving life and their connection with one another, The Oneness and God through acknowledgement of the birth of the Aquarian Age under the supervision of Archangel Haniel, who is yet to be recognized on Earth.

During the last century, the Aquarian Age was generally accepted to be important. The Five Groups of Children with newly altered spiritual perceptions began to be born in preparation for this transformation. It should be noted that many pioneers, who are way ahead of the general populous were born during the last century in preparation of what was to come. It was generally very difficult for them to establish anything that would create a big change until the last ten years of 1990 – 2000. These pioneers have taught, led and shown insights into the Aquarian Age. Their personal journey has been outside the box and often very emotionally sad. Looking back through this entire century, there have been a variety of pioneers all over the world who have changed Mankind for the better. These pioneers were encoded with each of the new encodings of the five new groups of Mankind. They are the models of ways to be in our future.

Those born before 1925 were generally Star Babies. Their traditions were important to them as well as their older family members. Some of their babies were also born with this same DNA coding during World War II and as late as 1950 onwards. Each Generation of Star Children is influential in helping the world to become a better place to live in.

During the build up and beginning of World War I (1916 - 1920), the pioneer Indigo Children were beginning to be born. Their psychic skills and talents were considered extreme at that time. Star Babies and Indigo Babies had a difficult time integrating their Lower-self experiences and were often in separation about spiritual matters once they were old enough to think for themselves.

As always, history reveals that Mankind adapts and moves on with the creation of his tools and his development of consciousness. By 1970 many Indigo Babies had been born who presented a deep need to be accepted as psychics to the world in general. They appeared to have a direct connection with The Oneness and with God, though the Lower-self was often lost in illusions and pretence. The battle for existence continued as more and more Indigo Babies became parents themselves. A new strain of Indigo Children became apparent as they blatantly demonstrated their psychic skills in even more unusual ways in public despite many threats.

As a result of these Indigo Children, by 1995, the world was full of potential psychics who proclaimed themselves or others to be messianic leaders. It became apparent that the dark was building in Mankind and so more people reached for the 'light' of God. Consciousness shifted toward receiving emanations of light and with that came the awareness of the presence of Crystal Children who were mostly born during the early 1980's, though a few were born as early as the 1920's. These Crystal Children brought a new spiritual consciousness to the Earth. Their abilities to 'know' Mankind was uncanny and scary for those born during the mid 20th century. Many were ostracized, treated as insane and placed in mental homes. Their instinctive and correct ability to psychically know each person's history, emotions and mental state in the blink of an eye was and is still hard to accept, especially since they announce themselves to always know best. They are always right and always able to see future results or choices that can be made. They have begun their work, in a variety of ways, to lay foundation stones for a new breed of Mankind. Many were afraid of these children and today they are still considered to have personalities that act outside the box. Some can be over-reactive and often pushy in their quest to get thing done, especially when it involves their own self-importance and the work they are trying to do.

Some unusually psychic children born during World War II were pioneers in directing the world to be aware of unity. These children were seen to be multitalented and often ahead of their time in many ways. It was clear that these children were unable to fit into any previously conceived category. They were so unusual in their ways, they were often ostracized, but bit by bit, they chipped at their forefathers ideas and brought a new consciousness into being. By the year 2000, the world was ready to accept something new. In the late years of the last century it was revealed to me that the Liquid Crystal Child's entry into this world had created a turning point in our awareness for a need to teach our children to repair this Earth.

On a Spiritual level, the arrival of the Liquid Crystal Child was the first step in the evolution of every individual spirit to be able to consciously be aware of Ascension. We all now know that we will be taken back into the fold and into The Oneness. There is no more doubt about the existence of God and our need to return. Today, these children are the teachers of spiritual matters that embrace Ascension without judgment. These Liquid Crystal Children can often be misidentified as they seem to be so similar in various ways to the other four types of children. It is important for them to network and unify Mankind.

Throughout the 20th century, The Hero Child has been quietly sorting the other four groups out as they appeared and performed during major events on Earth. They were then and are now the peacemakers and balancers. Their journey has been a hard one with little or no thanks for the good deeds they have done. Often they have lost their way, but with the influence and interaction of the Star, Indigo, and Crystal Children who were evolving, they have developed a clear focus on how the world should be rejuvenated. Now, with the awakened introduction of the Liquid Crystal Child's influence in our world, their work is once again becoming harder. They will struggle to keep the balance between the old ways of the Star Child's Group and the new ways of the other groups, which is being made abundantly clear all over the world at this time of writing. Wars, political coups, revolutions, mineral disputes, agricultural abuses, scientific atrocities and more are on their list of things to sort out.

Discover Your Own Child Group

Sit and contemplate your life, look deeply behind every event to find your true essence. Leave behind the pros and cons of what you did. Forget the emotions that held you back in pain. Only when you are ready to surrender to your own inner truths will your encoded *Soul Structure* activate your DNA within your deep-subconscious to reveal your purpose and spiritual destiny along with your Child Group. When you are ready and able to handle responsibilities in a productive way for yourself and those you interact with, you will want to create a different lifestyle.

You may be confused about your child type at first, and flip back and forth between the groups, since all have a common link, and are psychologically and emotionally able to identify themselves with one another. This is part of the evolution of *Ascension* in preparation for *Assimilation* into The Oneness.

If you are still in doubt, then seek out an Indigo Child to give you a reading, or better still listen to the wisdom of the Liquid Crystal Child who will reveal your personality and character in detail. After that, it is left to you to do something with this information.

The good thing about knowing which group you belong to is that you will immediately meet more of your own kind. By linking up with your own group you will develop physical goals to improve your lifestyle. The negative aspect of your life may prompt you spiritually and emotionally to accept a need to let go of some of your old friends and family members who are not on the same wave length as you.

By settling into your new awareness, you will make your life much easier. Old unlearned lessons will arise again, but in a new form for you to rethink your life and to subsequently set yourself up in the right environment, working in a pleasing way, as well as giving yourself time to be creative and nurtured. By then you will be in the second cycle of your life. This cycle can vary at different ages in an individual's life depending on their own personal *Soul Goal*. Of course, the remainder of your *Soul Structure* will influence everything you do whether you do them with a positive attitude or not. *(See my book: **The Rejection Syndrome**)*

If you are able to take an astrological point of view about yourself, you will find that there is a strong influence from the outer planets to direct you towards understanding your *Ascended Soul Group*. Your inner planets will help you to see many aspects of your personality and character. The influence of Neptune to your natal Pluto and Saturn will provide a great deal of insight about your spirituality.

Over the years, you will have noticed how you have struggled to learn who you are and what is your rightful place in society. Your levels of acceptance and rejection have been based upon what occurred in our history. The Piscean Age has gone. The Aquarian Age is now. Changes are unconditional and inevitable. If you are a mother, or about to be a mother, your emotional negative history must be erased in order to be receptive to your child's need.

Chapter Ten

Integration Of The Five Groups

Once your baby is born, it will be necessary to teach him/her how to integrate with members of other groups. Since each child is very different, your baby will have a hard time trying to connect. You should acknowledge that no matter how hard you try to understand his/her point of view in any given moment during any lesson, that you may not make the connection correctly. There can be no blame or shame cast upon yourself or your husband and family members.

No one can communicate unless we feel first, then visualize, listen and share. The old way was to sit around a table and dialogue until a solution was found, often ignoring feelings. The new way is to 'feel' one another intuitively and then act accordingly with inner knowing as we share our mindsets. This is not an easy step for the 'elders' of this world. Even as a young mother, you will still find this step hard to take because most of your mental conditioning occurred approximately twenty years before your baby has arrived. That was a long time ago when things were different.

While we can see the changes in electronic equipment, the shapes of new houses, the speed of vehicles, and lights from space ships and satellites moving around the world; it is not so easy to see the emotional and mental changes of individuals as they happen. We all have a habit of remembering how we were and trying to stay the same. As long as we hold on to what was, there can be nothing new created. The Piscean Age blocked emotional outlets. The Aquarian Age relies on emotional interaction.

It is interesting to note that the Piscean Age was a water age, symbolic of emotions, which were controlled by states of mentality. Now The Aquarian Age, which is an air sign depicting intelligence, is controlled by states of emotion. The physical result is a complete about face of attitude and action in form. What we accepted before as beliefs that would never waver, we have now abandoned in favor of new physical dynamics that allow creativity to express our existence. The road was paved with many hopes and dreams that never manifested. Now, in this new age, we create it by hard work every step of the way.

The integration of these five groups of children is the only way to embrace our future. New things are popping up all the time, and as much as we may groan about it, we eventually adapt and use those things. Now, we have to do the same with our relationships. We can flit from person to person to share ourselves, but we cannot build co-dependent relationships in order to fix us up. We must each learn to be independent of one another, yet at the same time, in unity with everyone. This way, we will all get along together and can then rebuild our home on Earth.

Mixing the Groups is easy. Literally reach out and touch someone. Hug them, listen to them and share everything that you are in as many positive ways as possible. If you play sports, congratulate the loser and acknowledge how much pleasure they have given you in providing you with an opportunity to explore your own skills. This is, of course, a mirror image for them too.

Perhaps you find yourself with people who are creative. If so, enjoy watching them in action and in the mirror image, do those things yourself without self-judgment.

In a working environment where the pressure is on, bring a little 'light' into the place with a present for everyone such as a few cookies and a cup of your special brew that gives a breather for them to enjoy,. Spread your appreciation for those colleagues around and watch them interact with you as you share your energy and radiate photons (beams of light from your aura) across the room.

The above tips are just some of the things you can do to see the positive side of you in action and to reap the rewards of your effort. Never manipulate yourself or anyone else to do something that is not pleasing. God made us and gave us strength, love and endurance to continue our existence forever; so a few moments in sharing are not much of a task to carry out in the midst of your busy life.

If you meet a Hero Child, listen to how they compare the past with the future. Interact with your ideas and plans and watch them adapt and physically support you in return for your emotional and mental interest.

When a Star Child crosses your path, you will know it. They will push your mind into ancient times, talking of old ways and old mistakes that must never be made again. Their words of wisdom will guide you away from repeating past mistakes. Yet, at the same time, they will remind you of the skills and talents of those who have gone before you which are now available for you to do again to develop a bright future.

As you greet an Indigo Child, acknowledge that they are carrying a message for you which could change your life. Interact with them and help them to expand their skills. In this way you help the Spiritual Teacher within them to evolve.

Interacting with a Crystal Child is always fun. They will constantly dart from one thing to another. Let them awaken you to new perceptions and opportunities that will develop your spiritual growth, along with establishing better standards of living.

With the exception of ancient spiritual teachers born during WWII who are Liquid Crystal Children, most Liquid Crystal babies are still young and infantile with a general majority being around 8 years old. There are a few who were born during the last 12 years of the 20th Century, who are now expressing great insight and wisdom to their peers. These young men and women are the Aquarian Age pioneers of their group. All are making a very strong influence on their parents and family's outlook, especially if they become parents themselves. The young ones are constantly on the go, looking for new things to do. They are full of questions and are highly intellectual. Their deep feelings are not to be ignored. They enjoy the outdoors, creative artistic activities and are easily bored with academic subjects which they master in minutes, if they are interested.

Having explained all this, it is important to mention that none of these groupings of child coding is new. We have always incarnated in groups. It is simply that we have evolved over time to show a different side of our awareness. In ancient times, these five groups were just as active. They simply had a more limited experience and were, therefore, less interested in *Ascension*. Philosophers and religious groups gave them different names befitting the times. Every spirit that incarnated in those days was still in a pattern of *Descension* and *Separation* from God.

The arrival of these groups of children, in our time, has not been locked into a definite time or place or century. These children are simply the evolved state of spirits now manifested into form who will continue to populate our world over the next two thousand years or so. As that time passes, they will transcend Earthly consciousness to embrace God as an aspect of themselves, then being able to manifest all the things that we believe impossible today.

It should be noted that we humans use less than ten percent of our brain's full capacity, and with the evolvement of and acceptance of *The Oneness* within our abilities, there is a great deal of room for further programming and learning. These new capabilities will allow a spirit, when in embodiment, to express truth in a variety of ways not yet even imagined.

The integration of the five groups of children will ensure our human continuation despite natural disasters. As the incarnating number of each group swells or contracts, a major shift in spiritual consciousness is achieved. At this time of writing, we are in the expansive aspects of our evolution. Spirits are being born to die quickly. Their life and death experiences leave a lesson which points out our general lack of appreciation for life and personal spiritual awareness.

A few spirits are living longer lives. Their personal journey reflects the enormous possibilities of the everlasting life of a spirit as it evolves consistently throughout eternity, whatever that may be. The human body can rejuvenate to last much longer if the mind and heart are positively in agreement. The more balanced the heart and mind, the stronger the spirit becomes in the flesh so that healing constantly happens.

The ebb and flow of these five groups of children is limitless and unstable. Within each group are young, mature, old and ancient spirits who are working in *separation* and collectively to aid their *Spiritual Soul* to become whole again. All Fragments (spirits) are constantly concerned with unity no matter in which dimension they dwell. Every living creature is connected to *The Oneness* and to God. You are the epitome of Divine Love in Form. Your unborn child will reflect that within you. Together you can build a new world for both of you.

Chapter Eleven

Your Child's First Year

Developing personality traits

From the moment you hold your child in your arms, your baby is watching you. His/her eyes are alert, body sensitive and ears listening. Every word you make is a jumble of sounds that resonate throughout his/her entire body. He/she is without muscular defenses and control and is, therefore, continually vibrating. For them, this is like sitting in an earthquake every minute of your day. Small wonder the nerves of your child will develop agitation and fear. But, do not fret; this is necessary for the inner workings of the brain to develop mental awareness and to accept survival and form.

During your child's journey through the birth canal, your muscular contractions will present him/her with his/her first real awareness of physical touch. Before that, your child was cocooned in liquid. As your baby travels through the birth canal, the brain is responding with a spiritual order from the *Soul Structure* to develop the first glimmers of their encoded *Soul Fear(s)*. This fear immediately activates your child's will to move his/her body out of the birth canal and to stimulate independent bodily functions. By the time he/she is out of the birth canal, all five of their physical senses are operating, including their sensory perceptions controlled by his/her psychic senses (spirit mind and emotions). (See my book **Expanding Images**). Every sense is heightened. A baby can cry and go into shock with so much happening so quickly.

When the umbilical cord is cut, the spirit of your child will slip into its Lower-self consciousness, suppressing almost all spiritual memories of where they have been in 'Heaven' along with wisdom from his/her Past Lives. The conscious mind is left to fend for itself. Now the spirit of your child must learn, day by day, to know him/herself. Their spiritual wisdom, lessons and capabilities are firmly locked away within their deep-subconscious awaiting the day when it will be appropriate to reveal those inner truths to him/her and others.

With the *Soul fear(s)* active, and the Spirit-self now primarily controlling itself through the five psychic senses of Psychometry (physical and emotional feelings), Clairvoyance (sight), Clairaudience (hearing), Clairsentience (smell and taste), a child will immediately stimulate his/her primary *Attitude* and *Modes* into play. The personality of your child is then manifested. Within hours, their *Goal* and *Archetype* will begin to show as *The Centers* are stimulated through your loving tender touch and deep nurturing with unconditional motherly love. (See Appendix). By the time your child is a day old, he/she has completely developed a personality and is truly in character. You may feel this, but be unable to notice it until physical, mental and emotional growth occurs.

No matter what your child's Spiritual Archetype Group, the many spirit fragments of their group are very close; they may be Angels, Spirit Guides or simply helpers. They will integrate their energy with your baby's energy to stimulate his/her *Soul lesson of Ascension*. At the same time, they will activate hidden spiritual aspects of the DNA strand, while bonding with your child's physical body. This will enable them to be in constant contact throughout your child's life. During his/her life, psychic abilities will provide strong links to them and *The Oneness*. Thus a link of unity is provided that will allow feedback over the years. It should be noted that this type of spiritual link has been the norm since the dawn of time on Earth.

By the end of the first day, you will notice how different this child is from any other children you may have had, even if the features are similar. Or, if it is your first, you may feel a sense of separation/loss and an emotional difference within yourself. It is not unusual for the first born to be compared with every member of the family. This is simply the way in which we rationalize our feelings and subtly identify a child's personality and character soon after birth. You should be aware that a mother's sense of separation often manifests outwardly through her own *Soul Fear(s)*. You may feel inadequate or unable to cope with separation to start with, but as the hours pass, your ability to physically cope will kick in nicely.

Recovering your own space is important. Many mothers forget about their own space during the first weeks of giving birth. They are so locked up in the routines that baby demands that an emotional depression can set in. It is important, that, as a new mother, you claim your space and rebuild your own spirit's strength along with rebuilding and nurturing your physical body. The more you take time to enjoy reading, walking, creating something for you, the better.

Since your baby has no words, or control over what is done to him/her, their main way of communication is to cry or sleep. Crying means "Help me." Sleep means "Leave me alone." As the days pass, you will be able to notice if your child is content or disturbed. Sometimes, disturbance is related to his/her own feelings of energy shifts and not to the physical needs of food or a diaper change. Often this can cause a new mother to fret unnecessarily. Just check your child and then quietly hum to calm him/her.

You Are Not Alone!

Sometimes your baby needs to cry in order to stimulate development of their will as well as to expand their lungs. If there seems to be nothing physically wrong, then look for energy shifts around you. Some of those energies may come from Earthbound Spirits who are seeking the 'light' and have sensed a new clean spirit in the body. They have come to get help. If you feel such a presence, then show them the 'light' through your own spiritual consciousness. Your baby's Spirit Guides together with yours will help them to leave by increasing the baby's aura along with your own. You both then generate a great deal of 'light' and heat. (Photons in action). Once an Earthbound Spirit accepts and is affected by the photons you both send out, their energy is transformed and they are ascended into a new dimension.

It is very common for new born babies to be a vehicle of transformation with the aid of their Spirit Guides to cleanse the world of lost spirit entities within the first few weeks of life. Unfortunately no one is educated about this and a mother's intuitive senses are often misunderstood as she becomes uneasy and panics with a supposition that there is something physically wrong with her child. It should be remembered that your baby has a wise self inside, as do you and that you both can and do cope with this kind of thing.

Earthbound Spirits are those who have lived a negative and fearful or tragic life. They have not been able to let go of their past life or been able to ascend into the Spirit World. They are literally in the 'Twilight Zone.' In the mirror image of this, your new baby is also in between two worlds. Their spirit self is still very much aware of its transformation and is now in a new space of consciousness. By the time a week has gone by, your baby will be fully present in his/her consciousness of this world. Any psychic senses that he/she manifests from then on, are signs for you to consider, which will eventually aid you in deciding which group of children they belong to.

If an Earthbound is very negative and carries physical and emotional trauma, you may well sense fear in yourself and your child. Simply go to a holy or sacred place and pray. Your Spirit Guides will rescue that lost spirit and bring them into the 'light.'

There are different vibrations of the 'light'. If a negative spirit is around you, you will feel hot and sweaty. So will your child. Simply hold your baby and think of the light; then visualize swirling rainbow colors all around you both. This will change your vibration so that the Earthbound Spirit cannot contact either of you again. When your energy is resonating at a different rate, only those who resonate in tune with you both can see and interact with you. A Spirit Guide or Guardian Angel will always resonate with you and help you. They will also rescue the Earthbound Spirit and take them to the waiting arms of their relatives in The Oneness.

Almost immediately after birth, your child will be aware of its own body and will respond to gas pains and various discomforts. Listen and check him/her out. If everything seems normal, but the crying continues, look inward for illness. If he/she is healthy, then look for strange energies in the room. It may be necessary to place your child elsewhere until he/she calms down. There is no knowing what type of fear is running though your child's undeveloped mind. All you can do is love them until it passes.

Depending on your child's coding from his/her *Soul Structure* and the group of children to which they belong, they may always be able to sense spirit people around them. In time, you will learn to know their ways and how to help them deal with their strange experiences. At no time, is your baby in danger. Every experience is part of his/her own personal growth pattern. They may seem innocent and helpless, but their spirit is strong and is active in any situation.

Independence

At three months, your baby will be establishing independence from you. This may not seem to be the case, as he/she is still so small, but you can notice interaction with other caretakers and acquaintances. Autonomy begins with social interaction. Your baby will be busy taking photo images and categorizing them into primary emotions that will be stored in the filing system of his/her brain. You have no way of knowing how that is done. You also have no way of knowing how you did it originally for yourself when you were born.

If your baby is showing signs of developing their own routine, then allow it to happen. If you invade his/her space, imposing your will, they will learn through their survival *Modes* to rebel against you later. Then you could have a temperamental child on your hands.

A sleeping baby needs exercise, so create a warm, but cooler atmosphere, which will cause him/her to wake up when it suits them. Kicking and screaming is good exercise. The more they kick, the greater the athlete!

By six months, your child should have a very inquisitive nature. Fill his/her waking hours with lots of bright toys with various textures to feel. During sleeping hours, play music; introduce new sounds that are nurturing as well as a tape of your voice. In this way, your child will not develop a fear of sudden sounds, but rather will embrace them as a stimulant to learn more.

From birth, avoid making loud noises like screaming and banging doors. This will frighten your child and cause him/her to develop phobias later in life. Arguing in front of your child is highly likely to develop a fear of social interaction with others, which will manifest inhibitions later in life. Never grab, shake or hit your child at any age. They have no consciousness of negative behavior and you would cause brain damage, so if the crying continues, go outside the front door and sit there to cool down while your baby cries away. Return when you feel you can cope. It should be noted that your child will mirror-image your mood, so if you are angry/sad, so will your child be. So, stay calm when being in close proximity to your child.

During normal activities, simply make faces that indicate good or bad. Give plenty of kisses and hugs to encourage development. Teach "yes" and "No" by looking directly into your child's eyes and saying the words calmly and firmly, but quietly. They will feel you with their psychic senses and understand. Repetition is important. Their brain is not remembering yesterday yet, but they will remember the feelings you give them.

Getting Around

At six-eight months old, your child is sitting and possibly rolling over, jumping up and down on his/her bottom to get around and touching everything in sight. Their curiosity is abundantly clear, so whatever toys are available to help them get around is wonderful. Your child must be encouraged to explore in order to develop social skills later. If you keep picking him/her up and carrying them everywhere, they will miss this important lesson.

Usually most children are crawling by nine/ten months and often seen staring into space. During those times, your child is looking at energy movement in the room. They do this with the use of their Third Eye Chakra. They may stare at the side of you, or at someone else. They are watching your aura flicker as you speak and do things. They may also be watching many Spirit Guides in the room. Every baby has the ability to see other spirits that are around them. It is not until they develop conscious understanding of words along with judgment that this ability fades away.

Some babies will find themselves able to stand and walk by 10 months. They are usually stimulated by their desire to touch everything that seems to attract them towards it. In truth, they are Old Spirits who are in earnest to get their life moving forward. They usually have very strong emotions and can get very angry if they find things too difficult. Their lesson of patience emerges from their coding of the *Chief Feature of Impatience* as would be showing noticeably a *Goal of Growth*. (See Appendix).

Some babies seem to crawl and hold onto things while very diligently stepping around carefully. Often they fall down on their bottoms and look scared, having experienced a nervous moment. They are dealing with their *Goal of Retardation* and a *Chief Feature of Self-deprecation* (See Appendix). Their little minds are busy exploring their inner *Soul Fears*. Their doubt is abundantly clear.

Perhaps your baby likes to climb! This type of child wants to see the world on your level. He/she constantly puts their hands up, asking to be picked up to head height. You may think they simply want a hug, but when you hold them close, they fidget and try to reach for something they have seen and want to play with. Children who do this are showing a *Chief Feature of Greed* along with a *Goal of Dominance*. (See Appendix). Once in your arms, they will be struggling to get down and touch the things they have noticed from above. Everything has a different point of view from your towering height.

Some babies just like to be different. They won't sit when you want them to. They run off, falling as they go, hurting self, only to repeat the same thing moments later. They seem to be constantly learning the hard way.

Their *Goal of Rejection* is paramount for this child in order to develop their will and freedom. Their *Chief Feature of Arrogance* (See Appendix) will push you to your limits. They will do what they want to do and not what you want them to do.

It may be difficult to outline the *Chief feature of Stubbornness* if your baby has a *Goal of Acceptance*. (See Appendix). They will appear to be in a world of their own. Nothing you say or do deters them. They will simply look at you, study you, and then return to what they were doing. If you take something from them, they will reach for something else. If they think someone needs something they have, they will give it to them without question. They do not cry unless punished severely. More often than not, they will appear moody.

If a baby is carrying the *Chief Feature of Self-deprecation* (see Appendix) they will immediately be very sensitive to everyone in their life. They will constantly mirror-image the goings on in the room. They will copy actions, emotions and try to adapt to mental energy. Often they will be confused and cry a lot. They will also laugh a lot when things go right. If he/she has a *Goal of Rejection*, (see Appendix) there will be times when they will not want to socialize and will grow angry. If an emotional button is pushed for lack of co-operation, they will demonstrate resistance in having to do something against their will.

A baby using the *Chief Feature of Self-destruction* (see Appendix) will pull everything he/she finds to pieces. They will bang and throw things around in an effort to discover their strengths. With a *Goal of Submission* a baby may well be in conflict both with everything they touch as well as with others who require emotional obedience. They learn that if they break something there are repercussions; the lesson of which is adaptability with awareness that nothing remains the same. This child will become frustrated from time to time with his/her inability to keep things the same. Ultimately, they must learn to go with the flow.

Often babies respond to the suffering of others. They watch and then run away to play. The child that has a *Chief Feature of Martyrdom* (see Appendix) will stay, climb all over you, get you a tissue and feed you his/her food. They are natural healers who, by their very existence, bring calmness into the home. If they are working with a *Goal of Stagnation*, (see Appendix) they will feel that there is no other place to be than in your arms. It will be hard to prevent them from clinging to your skirt or crying when you leave briefly. They will be fussy about trying new things and will demonstrate temper tantrums at the most inconvenient moment. They will manipulate you to put yourself on hold while you give them attention.

I have paired up the *Chief Features and Goals* to give you some idea of the things you can look for in your child. Since we can all choose a variety of these aspects in the *Soul Structure*, there are many possible combinations of how your child will behave. When we add the *Soul Fears*, there are even more ways for a child to handle life. See my book: **The Rejection Syndrome.**

Inbuilt Fears

Whatever your child's type and however many *Soul Fears* he/she has encoded into their *Soul Structure*, there are always ways for you to see them in action. Often these Soul Fears are acted out in the negative and seem to be the opposite of what is meant. These Fears are meant to be mastered throughout life. Here is a simple list of personality traits to look for:

Fear of Ascension: This child will consistently repeat the same activity until he/she has satisfied him/herself with an acceptable level of completion (mastery) in spite of displaying negative tantrums and frustrating yelling outbursts. He/she wants everything to be perfect and hates being bored. Their tantrums may seem as though they are giving up. All they want is attention and encouragement to continue trying.

Fear of Descension: This child will think nothing of pushing others away, keeping toys to self while consistently failing to use them personally or to interact with others. Sharing is difficult for him/her unless there is an alternative distraction. Competition is a constant battle in trying to be successful while worrying that they will be cast out socially and always fail. All they want is someone to do things for them.

Fear of Separation: This child will constantly be following others around while observing them from the edge of their playmates group; rarely joining in or remaining involved unless they feel secure. He/she likes to sit alone and be alone to do their thing. Later, after contemplation, they will copy others and chose their friends carefully. Social skills are often inadequate. All they want is a model to follow.

Fear of Assimilation: This child wants to be involved with everyone. He/she will do anything to get attention and to be in the center of the activity, even if he/she does not understand why or what is going on. This child will mimic easily and likes to entertain. However, they may panic when placed on center stage to perform. Most often, they will run away and hide for fear of losing control. All they want is comfort.

Fear of Divine Love: This child does not want to be swamped with kisses and hugs. He/she will run to you for hugs on his/her terms. Their mood swings from love to hate in the blink of an eye. They want you and they don't want you. They need help and they don't need help. He/she can be extremely controversial. It is hard for him/her to know what it is they really want, and even harder for those who love them. Watch out for manipulation of your emotions. All they want is to be in control.

By the time that first year is ended, your child is spiritually completely aware of their personality and is now ready to develop it with a focus on building character.

Chapter Twelve

The Second Year

Building Personality Traits

For every child, a battle of wills develops during their second year on Earth. Each child searches for their own identity while being taught to be something that everyone believes they need to be or to learn. Many of a child's primary behavior patterns are formulated during this year. They will show signs, at every turn, of their *Modes* (see Appendix) encoded in the *Soul Structure* to you. You may think of him/her as just being a child, doing childish things, but if you look behind the tantrums and the interactions which may be wonderful or destructive, you will see a bright little person with lots of personality who is desperately trying to prove him/her self to be as independent as possible.

Whichever group your child comes from, you will see their 'spiritual will' manifesting through their *Modes* as they begin to establish an awareness of their Spirit's journey in this life. With the combined connection of the *Soul Fears, Chief Feature(s)* and *Goal*, a child learns to communicate his/her feelings with words that correlate with their experiences. The words "Yes!" or "No!" when spoken out aloud can be interpreted in a thousand and one ways, depending of the way your child hears, sees, and feels you.

It would take an eternity to try to explain how the many ways the *Modes* can interact on emotional, mental and physicals levels of awareness. It is best that I simply give a couple of examples:

Example 1

A baby has a *Soul Fear of Ascension*, a *Chief Feature of Stubbornness*, A *Goal of Rejection* and the *Modes of Power* (mind) *Passion* (Emotions) and *Observation* (Body). (See Appendix).

Scenario

This female child could have had a cesarean birth, be lazy and resistant to feeding in the first few weeks after birth. *(Stubbornness.)* She could be constantly fussing for attention when feeling insecure, but is able to lie in the crib for an hour or so without needing attention. *(Rejection).* She will manipulate her parents to give her what she wants as soon as she realizes there is a way to manipulate. *(Power).* She will give plenty of kisses and hugs and earn merits easily. However, during these interactions with her parents and family, she will be observing them and learning how to get what she wants by becoming the Fixer Child. *(Passion).* By the time she is two years old, she will have learned to be the boss. Her understanding of trying to be in control will be well established. *(Observation).* She will spend the rest of her life testing herself and others. Her ways and means to stay on top can be delightful or despicable. She could be a living example of peace and happiness or a miserable, selfish and mean woman. Her choice of the way she interacts with others is the result of the way she thinks. Her thoughts are a result of her emotional experiences.

Example 2

A baby boy has a *Soul Fear of Assimilation*, with a *Chief Feature of Greed* and a *Goal of Dominance*, with the *Modes* of *Caution* (Mind) *Repression* (Emotions) and *Aggression* (Body).

Scenario

This baby, born naturally with an easy birth, could push himself into this world in a sudden unexpected early arrival. *(Chief Feature of Greed)*. His strong need to survive would stimulate his need to want more action *(Aggression)*. His aggressive desires and motivation for food and support will be uppermost, yet those demands for more will show discontent. He could keep his parents up all night, while they try to comfort him. *(Goal of Dominance)*. In the first six months of life, he will develop a great deal of reserve about himself as his *(Mode of Caution)* becomes more dominant.

A conflict will arise between what he wants and what he dares to get *(Repression)*. Even later in the second year, he will withhold his feelings and repress his growth. Some may say he is a slow developer. He could develop ADD traits having so much energy and no confidence to use it. His lesson in life would be to find his leadership abilities and skills to know when to act, and when to withdraw. No matter what his mood, his *(Goal of Dominance)* will ensure that everyone pays attention to him. As he matures, he will either become passive aggressive or outwardly pushy, laying down laws that must be followed to suit his comfort.

Developing Autonomy is hard for any baby. It is even harder for a mother to realize that her baby is showing signs of not needing her all the time. During this developing year, a baby should be encouraged to discover talents and skills and to be praised and encouraged to do more. If you let him/her be lazy, you will pay for it later. Allowing your child to wash, dress, and feed self is a very important step in becoming aware of their abilities. If you do everything for them, then they may not thank you later. A doting mother can also spoil a child's courage and desire to explore. So, allow your baby to touch, fall and explore as much as is safe. Most babies have a built-in awareness of what is safe, and with proper guidance will understand how to use it by the time they are 2 years old.

During this time, your child should be encouraged to explore their talents. Singing, building things, drawing and the like activities will assist their personalities to develop. Also, allowing your child to be with family members while you are out will teach them to not hang on your every word or action. The more you can educate your child to not need you, the more they will love you. Unfortunately, many mothers seem to feel that their child is helpless without them, and as the years pass into adulthood, they cannot let go. Then a child develops strong co-dependent relationships that may be very difficult to deal with.

Social interactions with games that allow your child to be swung in the air, dropped almost to the floor, rolled over, tickled and jostled will help him/her to overcome fears of physical harm. They will learn to enjoy a variety of active sports later. Their *Modes* will define what is safe for their personality.

Throughout your child's life, you will notice how the character will mature and evolve through the use of their *Modes* to develop a strong personality which may or may not be wonderful. During this second year, both you and your families will help develop personality traits according to the way you dialogue, interact and demonstrate. The more positive and nurturing you all are the better.

During this year, your child will be activating encoded history from your ancestors. You may see talents and skills manifesting that are not seen in your immediate family. You may also find that your child has physical DNA coding that can be a throwback to someone who is now deceased, but those who are old enough to remember them will see the likeness. Sometimes, a child will develop personality differences that seem to not fit in with the current family traits. This, in itself, is a good test for a spirit to learn to conform or rebel according to their *Soul Structure's* coding while on Earth. In the years to come, this child will feel different and unique.

Since all of the groups of children being born today are very psychic and talented, they will grow up very quickly. Any effort to hold them back will greatly upset them. They have a common passion to find unity through interaction and will not take lightly any punishments that seem to be ill-befitting their will. They will rebel if you are unjust or without understanding. So, even though they are only approaching 2 years old, they will let you know if they disagree with you.

Chapter Thirteen

The Next Three Years

The Formative Years

Life is full of excitement. Every day is a time to explore, learn and grow. Over the past two years, the child has mastered walking and talking along with the subsequent events that have stimulated your child's mind to begin questioning everything in existence, while accepting him/her self completely..

At this point in awareness, the spiritual *Centers* (see Appendix) of the *Soul Structure* will become more dominantly obvious. They will show their uncanny ability to 'know' what is right for them to do. Choosing comes easy, as does their absorption of intellectual data. Everything they learn is accepted as fact. Should something turn out to be false, they are devastated and easily hurt, but seem to soon get over it. Their primary coding for survival is pushing them to share, no matter how or who they do it with. They will be able to easily trust others, but unlike their forefathers, their psyche/intuition allows them to know if someone is untrustworthy. Then they will quite naturally withdraw to hide behind someone or to do something that takes them away from that person.

As social and behavioral skills evolve, each of the children in all five groups will learn how to interact with one another. The Hero Child will readily show off skills without being asked, while the Indigo Child will wait until asked before demonstrating what they can do in a very matter-of-fact way.

The Star Child will always like to do something with anyone, as long as they can tell everyone how to do it first. The Crystal Child will flip between each of the groups, exchanging energy and actions, but never really settling down into any special way. Lastly, the Liquid Crystal Child will be extremely sensitive to his/her surroundings. They will integrate with the other four groups when there seems to be an opening for them to show what they too can do. Until then, they patiently wait, watching and learning vicariously.

If you put one child from each of these Five Groups into a room, they will first watch one another; then the Star Child will do something, followed by the Crystal Child becoming involved. The Indigo Child will then decide to become the arbitrator between those two. Conflict or union can now occur. At this point, the Hero Child will try to take over. During this whole time the Liquid Crystal Child is weighing up the situation and may well distract the other four by doing something very individual which could cause them all to shift their point of view and join in with this fifth child. Watching two year old children's interactions can be very insightful.

By the time they are three years old, they will have developed a great deal of awareness about their individual appearance and subsequent skills. They will have formulated opinions founded on lessons learned. They will have taken on the moods, actions and beliefs of their parents, family members and teachers. Yet, despite this, they are uniquely different from their parent's generation. They have minds that see everything differently. They have a unique skill to skip into the overview; to see around any given situation. Parents will see their children as highly gifted and amazing individuals or extremely disruptive and annoying.

Because the *Centers* (See Appendix) of the *Soul Structure* are operating in every child born recently, it is easy to see how aware they are; knowing that what they do affects others, they will accordingly plan each next step. Conversation and discussion develop quickly. Their minds are supercharged with ability to take everything into their brain.

The conscious mind accepts everything as relative to a child's existence. So, to them, everything is important. A parent may not realize the significance of a child's thoughts. Sometimes, the actions and expressions of children are passed over as just a 'moment' which will be outgrown, forgotten and, therefore, unimportant. This was typically the way of dealing with children during the Piscean Age; but not so for children of the Aquarian Age. Their era is all about being in the now and from their point of view, all seems vitally important. They want to make an impression from the moment they are born. Even the chemicals in their brains are different. They can work out problems before their parents can even begin to see there is a problem.

All these children have a 'computer' mind. Everything that is seen, heard and felt by the conscious mind is passed into the subconscious mind where it is categorized into negative and positive groups, relative to your child's coding from his/her *Soul Structure*. From there, information is passed into the deep-subconscious mind, where everything is understood psychically. Their own spiritual interpretation of things that happen to them is very unique for them. Generally they are able to develop conscious memory at a much earlier age than past generations.

Consequently, a parent cannot just tell their child that what is occurring within the family or out in the world has no bearing on him/her and to spend time in their room. They will want to know and will push to know. Questions! Questions! Sometime parents do not know the answers, but usually their child does. Their spiritual self has awareness and a unique way of bringing forward inbuilt coding from The Oneness or their previous lives. They consciously remember their personal history and their interactions with others in history through visual images that arise from their own deep-subconscious mind.

Is Your Child More Physical Or Intellectual?

Every parent has a dream of their child growing up to be a success. Whether it is a doctor, lawyer or business entrepreneur, a parent projects their ideas of what success is. Unfortunately, for a parent, their child has already developed a primary program that has been stimulated in the womb.

Stimulation comes from either the Crown or Root Chakra at birth. This primary stimulation is what eventually evolves into the development of character.

Children that are encoded with an active Crown Chakra will develop their minds, stimulated by their spiritual coding from his/her *Soul Structure*. This will lead to the development of interests in more academic study such as technology and science, along with inspirational psychic senses. Those children who have a *Soul Structure* that has been stimulated by the Root Chakra will learn through physical activities rather than by observation and intellect. These children emotionally and physically feel their way through life and are likely to become farmers, animal and human caregivers, sportsmen, tradesmen and artists. They wear their emotions on their sleeve and let everyone become involved in them in order to mentally learn something about self.

This is not to say that they cannot do both. Some intellectual children are born with *Modes* that will awaken creativity within their minds to learn by touching and by trial and error. They may be hard on themselves, but will keep pushing for success. Meanwhile those children born with a physical need to touch will automatically develop talents and skills that come easily. When they master those skills, they may well develop other interests that can result in their being multi-talented. They could later, change careers and become more intellectual given the right stimuli.

The *Crown Chakra* encompasses the brain and overlaps the *Third Eye Chakra*. A child with a *Moving Center* (see Appendix) in the Crown Chakra will be extremely connected to their *Spirit Higher Mind Self*. As the *Soul Structure* is activated by outside forces, they will question their life's purpose and existence. Later in life, they may choose to ignore their education and family conditions by going on a personal quest to find out what their own truth is. They will be highly psychic and intelligent; always moving on from one thing to another. At the same time, they will have a great association with God, Spirit Guides and The Oneness. Their *Moving Center* will ensure that they never get 'stuck in a rut.' They will always move on when something does not feel right in order to become more independent. Despite their constant concerns for unraveling truth, they will search through logic to find deep emotions which can often confuse them and lead them down the wrong path. In this way, they learn hard lessons, but ultimately become stronger and more powerful.

Should a child have chosen to focus on the *Root Chakra*, sometimes referred to as the *Base Chakra*, everything they do will lead to an ever-increasing involvement in creating emotional ties through sexual activity and physical actions. In this way, the Lower-self beliefs that have been implanted during education will be tested. It is not unusual for a child with this *Root Chakra* coding called the *Sexual Center*, to use the *Spiritual Higher Emotional Center* (see Appendix) within their *Soul Structure* to stimulate their physical mind to search for situations that will involve a commitment to complete a lesson. The more involved the lesson, the more weighed down that individual child will feel. Emotions combined with rationality can cause many ups and down which often lead to a child building dependent relationships that require him/her to constantly seek approval and emotional support. Through involvement with emotional ties, the spirit of a child will sift through all their experiences and reasoned thoughts to find emotional truths. Those truths arise from deep feelings that have been stimulated to flow from their spirit self (Deep subconscious mind). Ultimately, through habit and fear they will become 'rooted' in their lifestyle and the people around them. Movement and change does not come easily, but spiritual growth can be a major awareness as Psychometry is used on a daily basis to feel and understand everyone.

Of course, it is possible to have both a *Higher Intellectual* and *Higher Emotional Center* with a sexual *Root Chakra* connection. These babies will be extremely psychic; emotionally very deep and in great need of close contact with those he/she loves. Or conversely, they could have normal *Lower Self Intellectual* and *Emotional Centers* encoded into their *Soul Structure*. (See Appendix) These children will not be very psychic, but will have great savvy about the people they love and the things that they do.

If they have chosen to mix up the *Higher Intellectual & Lower Emotional Centers*, they will be very resistant to common sense, being directed by their *Soul Structure* to follow their heart. Or, alternatively, they could have a *Higher Emotional & Lower Intellectual Center* which would allow them to search for good reasons to act, along with dishing out plenty of judgment.

In the beginning of the first three years of a child's life, these *Centers* are forming and developing their spiritual character on Earth. He/she will largely depend on the parents to help them believe in their own ways and trust their own heart.

Whatever the coding of the *Centers*, your child will be climbing up to the heights of elation and falling into the depths of despair. Their mental and emotional self will be searching to find a balance, learning to live in society, while searching inwardly for their true spiritual lessons.

Chapter Fourteen

Cycles Of Life

Higher And Lower Self-Attitudes

Everyone must have at least two cycles of life. This means that we each have periods in life as we age that are predominantly focused on the negative or positive results. Some chose to begin life in a negative cycle to learn the hard way, while others prefer a positive beginning. Later in life, through various circumstance that cause change, this cycle will adapt to the opposite. For Older Souls, a third and forth cycle are possible. Cycles of change are controlled by the coding of *The Attitudes*. (See Appendix). The result of adhering to any *Spiritual Attitude* is that it causes a number of years to pass by during which time a single way of seeing life is the focus. Collectively that number of years is called a cycle.

If we think of an oak tree and an acorn, we remember that the seed does not fall far from the tree. That little seed will grow beneath its parent tree until it has to fight its way upward, but in time the soil may not support them both, and one may die. This simple analogy is true of every human. When a child is born, the coding of the mother's *Soul Structure* carries all her fears, worries, anxieties as well as her hopes, dreams, talents and skills. Her small baby is encoded within his/her *Soul Structure* to be just the same as her for awhile. Their DNA strands are similar, but their cycle of life's purpose can be very different.

These cycles of life are created subliminally by the child's coding of their *Soul Structure*. Everyone has a dominant *Primary Spiritual Attitude* within the coding that ensures spiritual growth throughout life. They also use one or two other *Spiritual Attitudes* that manifest later in life to transform and accept changes of heart and mind. Conversation and repeated training will manifest conditioned everyday attitudes that arise from logical thinking. These thoughts have nothing to do with the coding of the *Spiritual Attitudes* from a spiritual point of view.

If a mother is in her first or second cycle of life and is insecure during her pregnancy, then her child will be born insecure. Her outlook on life will be adopted by her baby, who will copy her every word, expression and moods The *Primary Spiritual Attitude* will be subtly working throughout a child's life. But, in the first few years, he/she will take on the current *Spiritual Attitude* of his/her Mother.

There are seven *Spiritual Attitudes* available for each of us to choose to work with, but only the *Primary Spiritual Attitude* encoded within the *Soul Structure* will ultimately drive a person to find inner peace and truth. The seven *Spiritual Attitudes* are: *Idealist, Skeptic, Spiritualist, Stoic, Realist, Cynic and Pragmatist.* Anyone of these can be chosen to be the Primary one. The remaining *Spiritual Attitudes* will support the *Primary* one later in life when physical separation from mother is achieved. This can be during the teenager years.

Spiritual Attitudes are encoded to give each of us a general way to develop our own unique outlook on the world. By using the encoded *Spiritual Attitude* in each cycle on a daily basis, we each make up our own minds about how we wish to act. These *Spiritual Attitudes* should not be confused with attitudes of the mind created during an event where judgment controls the issue.

If the *Primary Attitude* coding of a mother is *Skeptic*, and her child's is *Idealist*, there will be conflict in the way they present themselves to one another. If the Mother is in her second cycle which has been chosen for her learning, such as a *Realist*, then all three *Spiritual Attitudes* will entwine together during her child's learning curve. So the combination of *Skeptic,*

Idealist and *Realist*, in this case, will cause her child to be suspicious about everything that she says.

He/she will need to investigate for him/herself and while searching for something that is different, cause him/herself to be distracted from one thing to another in an ever evolving quest to find the ideal. Consciously, this child will be learning to judge everything he/she sees as mother explains her opinions and enforces her rules concerning what is right and wrong. This may well conflict with the inner workings of the child's mind, especially if those thoughts and emotions come from a spiritual point of view.

It is hard to define each individual's *Primary Spiritual Attitude,* because the adopted *Spiritual Attitude* of their mother is often dominant, depending on circumstances and beliefs. A child may remain under the influence of his/her mother's *Spiritual Attitude* until they become an independent adult. At that time they will choose to develop a new way of looking at life and will attune to their own *Primary Spiritual Attitude* which has remained dormant until then. When this occurs, a second *Spiritual Attitude* may arise almost immediately, which will cause a big shift in their perception of self and the way they live. Should this happen during teenage years, a child is likely to drift away both emotionally and mentally from home life and desire to move out early.

This second *Spiritual Attitude* has a dominant effect on the Lower-self, (Everyday consciousness) often creating a false sense of self, because it is built upon innocence and the imposed perceptions and observations of others. The result is a conflict in beliefs that will force his/her personality to develop a deeper character. Consequently they will require support in; developing a need to make new friends, embrace the ideas of sharing and marriage, gaining and learning to appreciate belongings and in discovering activities that will encourage them to branch out with new ideas.

Over the next twenty years or so, this new *Spiritual Attitude* will strongly affect the Lower-self, controlling their conscious mind and their way of seeing the world; but underneath this will be their encoded *Primary Spiritual Attitude,* questioning everything they do.

The combination of the child's *Primary* and *Secondary Spiritual Attitudes* along with the daily Lower-self attitudes will often conflict with his/her Mother's encoded *Second Spiritual Attitude* (her second cycle) which was absorbed by her baby during his/her time spent in the womb. These three states of *Spiritual Attitudes,* when combined, will help an individual child to grow inwardly as well as to develop their own points of view along with ways to act in this world. Each child will face fear, pain, anger, guilt, loneliness, joy, pleasure, beauty and love as they flip their mind to suit every situation. While their rational mind may say they know their reasons, their *Spiritual Attitudes* are working subliminally.

When this cycle comes to an end, each child will become a mature adult who will be working on erasing their mother's *Primary and Second Spiritual Attitudes*, whether conscious or not. To do this they will awaken to their own *Second Spiritual Attitude* that has been deeply encoded into their *Soul Structure*. They will spend the rest of their life, working on themselves, emotionally, mentally and spiritually to find the best way to live a fulfilled life and to come to know God through their psychic intuition. By the time they pass back into the Spirit World, they will be feeling blessed that they have had a long life full of wonderful experiences that have caused them to evolve. Even their negative experiences will be perceived as great opportunities that have helped them to understand their life and the things they did.

The combination of all seven of these *Spiritual Attitudes* can be seen in your character. We all learn how to copy others and to mimic the different emotional and mental ways to express ourselves from them, so it could seem confusing. It is, however, possible to isolate the *Primary Spiritual Attitudes* of a mother and child when they appear to be in extreme support or separation. The remainder of the time, you will see their 'psychological attitudes' being flipped from mood to mood according to circumstances as they occur. But, if you look behind the issues, on a deeper level, you will see the *Primary and Secondary Spiritual Attitudes* working to create a personal cycle of history that may well be negatively expressed in the first and second cycles.

By watching your own mother and grandmother, you may well see some aspects of their *Spiritual Attitudes* that you have adopted while in the womb that were developed once you were born. When you see them, you will begin to also see them manifesting in your own child. These *Spiritual Attitudes* are frequently passed down from generation to generation. Each person must continue the lessons they agreed upon before coming into embodiment. Sometimes that lesson is to break away from old mindsets to begin afresh. The common focus is often to be free of conflict and to feel safe individually. A person can be then seen as a rebel, a misfit in the family. But, given time they prove themselves to be highly independent.

Once you realize how much you have been influenced by your ancestors' own *Spiritual Attitudes*, you begin to demand separation from those habits and misunderstandings that have caused you to lose your individual identity. A desire to find yourself during adulthood normally stimulates the development of your *Second Spiritual Attitude*, causing you to seek your own direction and purpose. So, be prepared to become a rebel who must find a cause that seems worthwhile to you.

Each *Spiritual Attitude* provides an opportunity to create short and long-term cycles of years according to one's free will. No length of a cycle is pre-determined. There will always be two or more cycles depending on the age of the spirit that is incarnated. Each cycle will be controlled by a single *Spiritual Attitude* that will jar a person to physically, mentally and emotionally tap into their *Primary Spiritual Attitude* no matter who you are or what they are doing.

The older a spirit, the more cycles there may be. A third cycle is often a time of contemplation and withdrawal or the development of a spiritual lesson that encourages self-awareness to share spiritual matters. During that time all Lower-self attitudes will be dissolved, having been used and learned from and then released.

The fourth cycle is usually artistic and creative. Sir Winston Churchill spent his last days painting during in his fourth cycle. No life is expected to end suddenly. If unforeseen events do shorten a life, then these encoded *Spiritual Attitudes* will be processed in the Spirit World accordingly. Everyone should find time to be creative throughout their life anyway.

It is only by contemplating your life as you mature, that you will come to understand which cycle you are in. The one thing you can be sure of is that you will always know when one cycle ends and another begins. You will feel it, see radical changes going on around you and definitely feel as though you have gone through a decided change in your understandings of the way you see yourself and the things you do. The trick in making this transference in energy change and spiritual awareness is to go with the flow, rather than to fight and resist change. Let go of material things that cause pain. Be open to new opportunities and always study to expand your mind and develop your spiritual awareness. Then you will be a good model for your child.

Chapter Fifteen

Awkward Growing Years

The Early Years

From three to ten years old, all children experience many bio-chemical changes in their bodies, as well as a multitude of negative and positive emotional events that cause great turmoil with their minds. According to the way he/she perceives their world, the brain has been absorbing information at an extremely rapid rate. They each have a strong awareness of who they are and are beginning to think for themselves. They have assimilated the many people they have met and are now becoming aware that their mothers are not perfect after all.

At the same time, they are beginning to recognize a need to identify on deeper levels with their father, grandfather, or some male role-model and, as a result, copy him in order to establish a separation from their mother's apron strings. Daughters want to be Daddy's little angel, while sons want to be just like their Dad, i.e., big and strong. Resistance to mother's encoded *Soul Structured Attitudes* or her adopted Lower-self attitudes set in. Mother begins to feel isolated and rejected as her child withdraws from her. She uses statements such as "You are just like your father!" regularly as she tries to cope with her child's lack of discipline or understanding from her point of view.

As your child sifts through all the data that has been accumulated from you and kindergarten to first year of high school, they have begun to understand that their life can be their own. However, with so many rules and regulations to follow, it is hard for him/her to know just how far they can push the envelope in exploring their independence. You may feel that your child is not paying attention to you, or not understanding what you are saying. The truth is, they are looking for another way to do things that will stimulate them to know more about their own spirit. If they remain tied to your ideas and to copy you, they will never achieve independence or awaken to their own spiritual journey.

Much of what your child learns in these years is cross-referenced in the subconscious and stored away for possible future consideration, should circumstances require them to awaken those memories. All their visual images are storing thousands of words attached to each image. Millions of images are recorded throughout his/her life. By the time your child has reached adolescence they will already have hundreds of thousands of images that may or may not mean something to them. Most of us never use many of those images on a conscious level again once adulthood is achieved. However, from a spiritual point of view, your child's deep-subconscious is assimilating all their subtle memories with present day activities.

Here is an example of how you may have reacted to your deep-subconscious when you were a child: At age two you eat strawberry ice cream for the first time. The sudden cold feeling on your tongue causes a shock wave throughout your body which leaves a slight feeling of fear. In later years, you seem to still love strawberry ice cream but an unexplained sense of uneasiness follows each time you eat it that is usually rationalized away as related to a current condition or situation.

The Adolescent Years

As your child develops awareness about him/herself, the egotistical self begins to develop. They will strongly express opinions as you give your demands for obedience. Their resistance to the way you say something and how you look will be uppermost in their mind as they respond to you. For example, you could be offering something wonderful, but in their mind it could be boring or simply something that is too new for them to deal with. As they begin to pay more attention to what they want to do, their conscious mind begins to slow down. Suddenly, learning is not so easy. The pressure of being a 'super' child begins to show.

If yours is a Hero Child, they will begin to develop their own ideas about what is good for them and when to do whatever they decide to do. In some way, they will find the means to get out and about despite any restrictions you may apply. They can be dishonest, telling you one thing while planning to do another. They may do some things you ask them to do while hoping to get away with something else later. In some cases, they will argue with you to try and convince you that their ideas are the best.

If your child is a Star Child, he/she will approach you directly and tell you what they want to do and expect you to agree. If you don't, they will persist in asking "Why," so that no matter what you say, there is never an answer that is good enough. They will eventually give up, but not before letting you know in no uncertain terms that you have not had the last word. They will be back with more demands by day's end. They are likely to stomp around, be moody or leave the house to sit outside, waiting for you to come and find out what is wrong.

If your child is an Indigo Child, he/she will actually try to nurture you by providing you with information that he/she believes will put you in a good mood so that when the big request comes from them, they expect you to agree and give them what they want. If you deny them, emotional outbursts, stomping and yelling are likely to follow. They will use verbal words of hurt such as "I hate you! You never do what I want!" Moments later, they will ask again for something else. They just want your attention and the attention of others.

If your child is a Crystal Child, he/she will be impatient, wanting everything to happen immediately. When they ask for something, they expect an immediate response that will give them an opportunity to escape a predicament, such as doing chores or homework. They may well be frustrated and on the go, searching for something new to entertain themselves, and if you refuse to join in, then you are neglecting them in his/her opinion. Tantrum behavior will follow.

If your child is a Liquid Crystal Child, you will discover that nothing really keeps them happy for long, unless they have an inborn talent that appears at an early age, such as playing the piano. They learn very quickly and you will be demanded to pay attention to their every whim. Your lessons in showing patience will teach them to slow down and learn to enjoy one thing at a time. They will also reflect your moods; so stay happy and sparkly, avoiding over-exuberance which results in hyper activity that will wear you out. This type of child needs discipline and exercise to train the mind to focus. If there are no luring activities, this child can be very destructive and annoying. Set rules and keep to them no matter what they do to get you to change your mind.

Puberty And Emotional Growth

The bio-chemical changes in your child's body as their reproductive organs mature are enormous. Outwardly, the conscious mind notes the changes and tries to adapt mentally and emotionally. However, the chemical changes that occur in their body are erratic and hard to deal with. Sleeping patterns change, as do psychological attitudes. At this time, the Spirit-self is awakening to its own ability to develop independence in new ways. At the same time, their brain is reducing many active cells as they condense information into a more solid personality and character. As a result learning is slowed down to a more normal pace.

Since they are still a child in your eyes, their inability to successfully maximize their skills, talents and abilities will worry you. Social contacts, outside influences far beyond your awareness, as well as all their teachers, religious advisors and the like, including TV, movies and media, will affect their behavior even as you try to control it. The only way you can identify with your child's mind is by comparing your own history and the things you went through. Unfortunately, those are your own points of reference and not your child's, so no amount of explaining and identifying with them will work. Every new generation will tell those who have gone before that "Things are different now!" Indeed they are!

Your child has a different *Soul Structure* with a different focus from yours, even if there are similar traits. Your technological training is different from theirs. Your skills and talents may be undeveloped compared with theirs, or they may have no interest in the things you like. You are indeed 'worlds apart.' Yet, despite this, you both know that there must be something in common that still binds you together. Those things are likely to be variations on talents and skills as well as the natural 'Laws of Karma,' which are God's rules to be obeyed.

I have spent a great deal of time in my other books talking about these laws. Simply put, you can never ignore them. They always arise to hit you in the face when you least expect it.

Chapter Sixteen

Watch Out For Negative Physical Indicators

Your Teenage Child

Sad eyes, downcast body language, lack of communication, withdrawal into their own room, or escaping outside are all indicative of your child's negative expression when they feel that the First Spiritual Law Of Karma has been broken: that of invasion into their own personal space which encompasses their spiritual space too. By repeatedly telling your child what to do according to your own ideas, you will push them over the edge, creating an emotional void between you. Likewise, if their peers and teachers are over-demanding with a lack of regard for sharing your child's point of view, they will cause your child to hide self away from society.

We are all familiar with teenage behavior patterns: constant refusal to do the chores, untidy behavior, lack of socially productive friends, bad grades, and uncooperative team spirit etc. When these types of things happen, a child is expressing their need to learn responsibility, governed by the Second Law Of Karma, that we each shall be responsible for all that we do. If you take charge too often, your child will respond negatively towards your efforts. They will not appreciate your sacrifices. Let them find their own way by failing and losing things. Nothing is more frustrating, but it does eventually lead them to organization and discipline. At school they may be positively active, making sure they are in control, while feeling inwardly out of control. Life is a big dramatic show and he/she must learn to deal with it.

The Third Law Of Karma is about integrating socially and spiritually. On Earth that is often done in places of worship, school or in sports, dance classes and the like. Here your child will learn to share and should enjoy these activities. If you force them to go to classes to become something that is not in their nature to want to do, they will rebel. There must be fun in the sharing and joy in the doing. The minute it becomes a chore it becomes a suffering. Your child could show signs of fear, anger and failure and then develop excuses to not go. They should not be forced to be something they are not ready to try.

Your baby is now becoming your teenager child who wants to do what she/he wants all the time. These growth years are often topsy-turvy and much of what comes to pass is simply a stepping stone leading them on in the right direction towards adulthood. Meanwhile, they moan and groan about everything during the act of sharing while doing something responsible. Your child will constantly search for ways to change your attitude and get you to adapt to their way of seeing things. It is important to remain firm but flexible and to be a model by sharing and being spiritually aware. They will find unity no matter what occurs.

The Fourth Law Of Karma ensures that we all realize just how similar we are. We attract those who are into the same things. During this time, it is important that your child learns to socialize and integrate while recognizing their own choices. To act foolishly, is to attract others who seem to be even more foolish, even though they may show it differently. Through these teenage years, children will congregate in groups around the school yards and playing fields. Their bravado will often be covering a great deal of insecurity. Their competitive skills are being tested along with their ability to make independent and wise choices. It is important to encourage your child to understand that the mirror image of what they see can be found inside self, even if it is only a thought. The idea of trying a drug is one step away from watching someone else take the drug. Loyalties get tested and the honor system should always be backed up with parental support. Set yourself as a good model. If you do not want your child to smoke, then don't smoke yourself.

The Fifth Law involves self-esteem: having a good sense of worth and values. If a child is educated to understand that God is omnipresent in everything he/she does, then that presence will inspire him/her to do the best that they can to succeed, even when up against the odds. If you teach your child to be aware of the power of their own spirit within their body to manifest the best that God has given them, they will expect the best and never settle for second best. God did not make us to create negativity. Each child must learn to see the best in everything, even when failure occurs. With your positive support, your child will acquire a good sense of harmony in all things and will feel the all encompassing Divine Love of God by understanding how the light and dark create balance. In other words, they learn to take the rough with the smooth and not judge self or others, but rather learn from it.

Always try to see these Laws Of Karma working and allow yourself to use them wisely. In this way, you will be a better parent and closer to God. If you do not obey these unwritten laws yourself, you can be sure that your child will disobey you. You could have anarchy on your hands.

We saw these Laws of God being constantly broken during the Piscean Age as Mankind invaded foreign lands, plundered and pillaged, murdered and manipulated one another in the name of God. In this Aquarian Age, we must learn to share and appreciate all that we are. This is the age of change and our children are here to manifest this change.

The Hero Children will step forward in school to save the weakest child from being victimized. They will lead the pack, by stimulating those who follow them to do the right thing. If they have been exposed to a great deal of negativity, they will fight to gain leadership for the benefit of the weak. Either way, they will learn to fix themselves and others. The Five Laws of Karma will constantly push them to settle old scores and make new plans that involve new ways that will save the day.

The Star Children will evaluate the other children in their classrooms, searching for those who have a good understanding of the rules and regulations. They will try to be peacemakers, often finding themselves torn between points of view. They will debate any issue in search of a well earned victory when change occurs. If exposed to negativity in school, they will find a teacher to sort out their problems, or turn to a best friend/counselor for help. You will be the last to know, as a parent. They want to solve their own problems away from family environment. They are capable of being secretive and wishful and can get lost when up against stronger children.

The Indigo Children will feel everyone in their class and will identify with them all. They will push themselves to understand how the others use their minds and hearts. These children are often very creative, imaginative and complex. Most of the Hero and Star Children will misunderstand them. The Indigo Children are psychically different. They use the Laws of Karma in spiritual ways, often feeling the fear, anger and worries of the other children. If over-exposed, they can become negative and traumatized by events that occur around or near them, even if not involved. These children will search for others like themselves, and stick together through the years, making lasting friendships. Parental guidance is vital.

The Crystal Children are often bullies and disruptive in the early school years, but later in life, settle down to routines that provide a more tranquil comfort zone. School friends come and go along with their activity choices. Their minds can be brilliant at some things and hopeless at others. Their need to integrate, touch and see everything heightens their curiosity to a point where they may intrude upon the space of others. They often forget what they started and fail to finish. These children need help understanding organization, scheduling and structure. Once learned, they become model citizens who never want to break the rules. But until then, they have little concern for rules. Left to their own devices, these children can be very destructive. To touch is to hit, to hit is to defend etc. They need to learn to be soft and gentle. Plenty of parental hugs are delightful and meaningful to them.

The Liquid Crystal Children are a delight. They seem to know about everything that you show or tell them. They learn quickly and don't understand why others are far behind them. They have a very high IQ and can often read, write and count by three years old. They develop skills and talents at such a fast rate that they can be called a genius. This can be their psychological downfall unless parental guidance is constant. Other children may be afraid of them, their skills and talents and want to test or destroy them. This can make a Liquid Crystal Child feel isolated and, as a result, unable to socialize with his/her peers, while getting along famously with adults. A lack of emotional understanding can create a large gap in their maturity and social education. Encourage this child to be an all-around student and to share and enjoy everything with those who would be friends with them. On the negative side, this type of child can judge and condemn others easily. They may quote history and simply come to high-minded opinions about themselves that will serve only to isolate them even more. Ostracism is the absolute and ultimate failure for this child.

All five types of children have their own special abilities. In every case, a child needs to have parental guidance along with spiritual awareness. This is a new generation of humans who have come to the Earth to experience Mankind in a new way. They will never settle for the old ways for the sake of the old ways. The Hero Child will explore new adventures and push him/herself to the edge of their physical abilities. The Star Child will question their very existence, compare the past and examine the possibilities for their future. They will invent what needs to be done. The Indigo Child will spiritually question the very nature of Earth, the religious rules and beliefs and develop a special bond with those who would build a better world to live in. While these three types of children are working together, the Crystal Child will question them all. His/her reason for existence will define the edges of creativity, as well as our technological growth, constantly searching for a better standard of living for all. The Crystal Liquid child will become the scientist of tomorrow. Their search for further understanding of the Human race, our DNA, our health issues, or discovery of new energies and the abilities of Mankind to evolve intellectually, will awaken a greater spiritual philosophy that will harmonize our purpose for existence – to become more like God. In their quest to develop further, they will recall historical past in detail, which will include the emotional, mental and physical aspects of any given situation. They are deeply affected and stimulated to move on by these encoded memories of by gone days.

The combination of these five types working together over the next hundred years will create a new world that is, at this time, beyond our understanding, just as our great-grandfathers could not have imagined the world we live in today.

Early Teenagers And Sexual Awakening

With awareness of others, and the development of the body, sexual expression becomes a vital aspect of social integration. Each type will be hopelessly lost in a sea of emotions concerning their identity and their abilities to become sexually active. Children often explore their bodies privately, as well as socially by interactive kissing, hugging and making eye contacts with everyone they feel safe with. This feeling of safety can often be misconstrued to mean sexual experimentation.

Since the media world is constantly bombarding these children with photos, movies and news emphasizing sexual activity, these children are quite naturally drawn to making early choices to explore themselves and others. Of course, girls get pregnant and boys suddenly find themselves facing responsibilities beyond their capability. Though abortion is available, most children today will opt for adoption, despite the popular view that abortion is better when a child is so young. While I personally do not have a point of view about this subject in everyday terms, I have fully shared my spiritual experiences in my book **Pro-Life, Pro-Choice, Pro-Spirit.** It is to this point that I now refer the five types of children.

The Hero Child will be daunted by the idea of being a sexual failure and will hold back from intercourse, but will not be afraid to experiment with plenty of kisses and hugs. If their partner is not compatible, they will be the first to admit it and break the relationship. If their heart is captured, they are likely to be very serious about their relationship and will expect the bond to deepen and last forever. Should a relationship be ended by their partner, they will be devastated and wonder why they were dumped. It will take them some time to recover and, during that time, they will be soul searching to find a better way to make a stronger and safer relationship next time. Should they have experienced intercourse, they will be very cautious about future intimacies and may develop a deep-seated fear of physical sharing, having convinced self that they are not good enough.

The Star Child will be happy to experiment all the way. Their idea of sharing is to explore their own feelings as much as possible. Commitments to a relationship will be controlled by circumstances that surround them. If it is not socially acceptable, they will secretly admire and long for someone who is unobtainable. Should there be a liaison, they are likely to keep their emotions private. When a relationship ends, they will judge and condemn their partner as unfaithful, unbalanced or useless and therefore, an unnecessary part of their life. They will quickly mend their broken heart and immediately look for someone else to love. In this way, this child is able to experience many aspects of self during several relationships that never last, but which all seem highly important at the time. In their minds they are searching for the perfect partner and will not be satisfied with anything less. However, they may well stay in an unhappy marriage/relationship for years because of a deep-seated fear of being alone which may far outweigh their misery.

The Indigo Child will be drawn to those who have a good sense of social graces. Their open heart will search for a partner who seems to understand them. They will seek solace and comfort in the arms of the one they love. Sexual sharing will not be as important as their conversations which will always focus on or around their lifestyles and their hopes and dreams for a good future. These Indigo Children want a relationship that lasts and will be emotionally devastated when their first love fades away. That person will forever be locked in their heart and will be the foundation stone for future relationships. It is hoped, therefore, that this first love is a good one. Since few Indigo Children will stay with their first love, it is important for them to experience both physical and emotional ties before sex. Physical sexual activity is their final way of sharing which can be sacrificial and that could have a profound effect on their mind if they are abused or misused from their point of view. Their hopes and dreams are dashed as Prince/Princess Charming ride off into the sunset without them.

Crystal Children are the heartbreakers. They will want to explore themselves and others in a variety of ways, including early sex at a very young age. Their intimacy levels are momentary and can flare up in great waves of emotion or cool down in the blink of an eye when something else or someone else comes along. They are constantly adding new information to their lists of wants when it comes to their relationships. They may need room to be independent or conversely, want their partner to wait on them hand and foot and will get angry if they do not get the attention they need. Their sexual activity is likely to be fast and furious if they are negative and can develop deep-seated insecurities about their abilities to satisfy their partner or self. Inhibition is highly likely following several love experiences where no emotional ties have been made. Those who are fortunate to find true love will control their relationship in a variety of ways to ensure it lasts. This can develop various control and clinging problems later in life.

The Liquid Crystal Child has yet to be understood in physical relationships. However, it is said by Spirit Guides that they will be enormously attracted to one another for their abilities to share emotionally, mentally, physically and spiritually. Once this type of bond occurs, they will move deeply into sexual activity as a way to express their needs and to satisfy the needs of their partner. Sex will be very important to them as will marriage and family ties. They will seek friends for life and will enjoy the many talents of those friends without judgment. Should a Liquid Crystal Child have their heart broken, they will become a recluse for a period of time, sinking their mind into some task that will allow them to rebuild their energy and clarify the lessons they have learned about him/her spiritual self. Once they have faced their issues, they will open up again to another deep-feeling relationship that will offer even more spiritual growth. This type of child will not be inhibited by sex, but will be hurt deeply by words and looks that can offend.

During the mid-late teens these children will integrate their energies. By crossing the sexual borders, they will learn more about one another, their types and their purpose. Each relationship these children make will have a lasting effect on the world as a whole. Their interactions will set the pace for future generations. During the Piscean Age, men and women had separate roles, clearly defined by their sex, society and class. In the Aquarian Age, men and women will be unconcerned about their sex, but extremely concerned in how the sexes integrate in a common bond to create a better future for all.

Chapter Seventeen

Late Teens Into Early Adulthood

No matter what the generation or the period in history, teenage years have always been an awkward time. Not still a child, yet not really an adult, each individual tries to fit in with society and integrate their personality and character with those they love, admire or simply dislike. Throughout history, men have modeled themselves on heroes of the time, while women have sought out strong, independent people to rely upon. During the last hundred years, there has been a *shift* in that consciousness to prevent the human race from carrying on in such a way. This is the beginning of the biggest shift of all time. It will not come easily, but like any teenager, the growth that each must go through is paramount to adulthood. We as very young spirits must learn to grow up and be part of The Oneness. Yet, in spite of this we are all uniquely different which makes us equal parts that make up the Whole that is The Creator. We must all accept that our journey takes us through earthly ways into Ascension and Oneness.

In the past, throughout many wars, social structures have been destroyed, giving rise to new ideas that have increased technological, social and economic standards. Our systems of survival have changed. Wars are fought differently. Countries have been tied together as a result of world trade. All these new ways have caused every nation to change their attitudes about men, women and children as well as the care and support of their animals. Once, women who were well-rounded and moderately attractive were admired by society. Now, we look for skinny, lanky women to demonstrate an ideal body. In the past, men who were intelligent and able to sustain a family standard of living were admired and expected to hide their problems from everyone. Now, men can show off their muscles, stay home and play with their children, which will then allow a wife to take on the responsibilities of work. Men are now expected to share deep emotions through all forms of interaction with family and friends.

This world as we know it is in danger of losing many species of animals, plants and trees as we enter more and more of the industrial age of plastic, computers, phones etc. We no longer see the beauty of our surroundings as we walk to work. Instead, we are planning business structures that cause us to peer into our computers for hours on end or have our minds set on conversations on our cell phones as we drive or ride the trains. Gone are the days of children's hour; when parents spent time with kids in parks, zoos and country settings. In their place are indoor game areas where parents can let their children run free with an idea to exhaust them quickly, so that bedtime comes quicker. Gone are the times of sharing conversation around the dining table. In its place are quick meals from plastic pouches, TV shows to watch while eating with no interaction with family members, other than to argue over who has the remote and what program to watch. Small wonder that we find children withdrawing to where there is no contact, hugs or love shared. We have reached down into the bottom of the pit to learn that *isolation is destructive*. Who knows what a child's mind is thinking when no one asks! Who knows what they miss when no one looks around! Who knows what goes unlearned when no one listens! Yet, despite this, children survive 'safely' within their own mental cocoons.

In the shift of this new Aquarian Age, the spiritual essence of each individual cries out to be seen, felt and shared in the purest way possible. Those who lack this kind of bonding and attention are full of jealousy, hatred and judgment against those who try to build a better world. All over the world, there is conflict about how we should integrate our species. Unfortunately, it will get worse before it can get better. We must wait for older generations to die off, their off-spring to fade away from control, and their children to build a new way of life. In this way, the young children born in the Aquarian Age will be able to establish a better world to live in. If they have their way in the next hundred years or so, we will have a world of peace as foretold.

In the over-view, The Oneness is inspiring us all to find a way to connect. We each are now learning to appreciate our thoughts, our emotions and the value of our life along with a positive outlook on our future. There is a great deal of interest in psychotherapy, psychology and natural healing skills. Whatever, the choice, there is always a change in an individual's perspective once these topics have been personally discovered and then explored further. It is only a matter of time, before the "hundredth monkey" causes our world to *shift* into a new way of sharing.

Each of the five kinds of children born in this century will be looking at their future. They will have firm ideas about what they envision and hope to learn and master quickly. In the late teen years, their ideas and actions will affect those who are their leaders. They will make a stand against anything that appears to be wrong to them. As soon as they are old enough to vote, they will. As soon as they are old enough to carry a weapon, they will. As soon as they are wise enough to teach, they will. So, it goes on. They are a force that cannot be stopped.

At the time of this writing, the world in general is at war. Everywhere, there are children who have been born into suffering, trials and tribulations of various kinds. In countries that are financially established, they are complaining about the systems of education and politics. In Third world countries, they are learning about judgment and loss. No matter how serious the situation, all children are aware of a deep sense of loss. It cannot be measured by things. This is a spiritual sense of loss. Children are learning to care about other children. Their teachers are of the Piscean Age, but those who lead are the pioneers of the Aquarian Age. They have survived, finding their own ways to fight the broken systems of the world. They have become the spiritual gurus, politicians and the like. Their personal journey has been to open a portal of energy for the children of our future.

Early Adult Life In This Century

Some of the Aquarian Age Hero Children are now adults. They are the offspring of the children of approximately the last thirty years of the Piscean

Age. They are influenced by what has gone before, but prepared for what will come to pass. They have strong political views about government, war, and education. They are also extremely interested in sports and social interaction. Many of these children are becoming teachers, social workers, professional business people, doctors and therapists. A Hero Child, born late in the twentieth century is now learning to establish self with a profession or skill. Others simply move from job to job in a journey of discovery of what it is they really want to do.

The Star Child, like the Hero Child, is the offspring of those born during the Piscean Age. These children are older now, but have been fighting for their rights in a variety of ways since adolescence. They like to debate, set a challenge and ask questions. In their working lives, they have become the professors, workhorses in production, laboring to build structures that will provide better living and working conditions. They are interested in the arts, music and creative pursuits that will leave something for the world to learn. Some of these children are in their late forties and fifties These Star Children have been the pioneers that crossed the barriers of our races. The Star Children who were born later are now establishing themselves in inspirational ways though technology, the arts, design and motion and are likely to find careers in media, written communications and computer education or within industry or educational cultural fields.

Many of the Indigo Children have been around since the early 1930's but were not called that or noticed for their abilities. They have been the soldiers who have quietly marched to their own drummer, while following along with the rest of society. They have in their own way, created a doorway through psychic senses, the arts, healing and education on a spiritual level. It has only been in the last twenty years that these children have been recognized for their natural talent in demonstrating spiritual psychic abilities in physical ways. Simply put, their spirit selves have evolved to a point of awareness of ascending in vibration and are able to manifest more abilities on Earth. With this point in mind, young Indigo Children who are now becoming teenagers are amazing in their ability to know the future. They have made up their minds about their careers which may vary immensely in style and approach as they evolve. They will walk through all sorts of similar life jobs as they grow spiritually and share themselves mentally and physically with others. These Indigo Children can read your mind, interpret your emotions and explain your insecurities. Whatever job they take, they will always be one step ahead of other types of children.

Some Crystal Children were born during the last twenty years of the last century and now many are young and very expressive, being driven by their curiosity. They abound with energy, love sports and physical interactions enabling them to socialize easily in both negative and positive situations. They will productively search for a career in law, or something that allows them to find the balance between what is good or bad. They will make excellent teachers, counselors and probation officers as well as becoming active in sales and marketing around the world. Their need to know other cultures and to experience different parts of the world will stimulate them to look for work abroad or to take holidays whenever possible in far-away places. They are constantly on the look-out for new things to do and new ways to be. Change is exciting to them as is the challenge of life. They are daredevils and often fearless, so, an interaction with others comes naturally in spite of any rules and regulations to the contrary. If a sign says "Silence," they will talk about it. Often they are rule breakers, but with a purpose to create a new rule. They may change their job focus many times.

A very few Liquid Crystal Children were born in the beginning of the last century and have throughout their lives, imparted their wisdom. Some who were born during WWII are still following in the footsteps of those gone before them. Their uncanny ability to know and adhere to ancient times where philosophy and doctrine was first accepted is paramount to their work. They have walked a lonely life, but have made sure that their work is recorded for prosperity in some form or another. None of their work will be forgotten.

Today the new Liquid Crystal Children being born have far reaching abilities to continue their teacher's works. They will be concerned with what has gone before and what has been learned and how to take the knowledge to the next step. They are the archeologists, the historians, and the record keepers. Their ability to keep financial and statistical records will lead them to jobs that challenge their minds and their memories. They learn extremely quickly and easily become bored if there is no new input. They will be interested in preparing or creating something new out of the old. For example, they will be interested in the DNA strands and their personal differences which will guide them to study this. These children have the record of all creation within their genes which can be tapped into at any given moment. They can excel in any kind of work, art or music. They will create whatever is needed for our future. They will instantly know what is wrong and correct it without consulting others. They are the pioneers of the future race of Mankind, so they must be able to do any kind of work in any kind of situation. These children are being born all over the world now and will be on the increase as the years pass.

Much of what I have written in this section may sound like history is being repeated, and so it is in many ways, for it is the very nature of Mankind to repeat cycles and to grow out of what was. The difference, this time, is that we are programmed to be awakened into spiritual consciousness. While this has happened in ages gone by for philosophical reasons, it is now happening again with a new point of view, that of Spiritual Ascension into the heart of God, whereas in the past Mankind has always been concerned with existence and the values of physical life on Earth. Now all eyes are turned to the Heavens and beyond.

Philosophers have speculated about birth, life, death, etc. and have often had insights which have later led to an awakening of the Lower-Self and an opening of the conscious mind to question why one exists. This time, it is not about physical life, but about spiritual life. Everyone will eventually be aware of their spirit within their body and will be open to the power of their own spiritual abilities. At this time of writing, there are those who think this power will be used for evil. In the Oneness, there is no evil, only innocence and ignorance, both of which are easily transformed into wisdom by embracing Universal Consciousness and Divine Love. We all have the Liquid Crystal Child codes within our spirit bodies where all of creation has been recorded within our spiritual vibration. For want of an expression we will call it the DNA of the spirit body. It is the first time that this spiritual coding of our history has been passed on down into the human form. On an everyday level, these Liquid Crystal Children will seek a spiritual rebellion. The dark will fight light within each one. Their Lower-self will mentally and physically fight their Spiritual Higher-self emotions and actions.

Chapter Eighteen

Religious And Spiritual Education

The First Spiritual Age

It is no surprise that religions of the world, supported by each of their own various indoctrinations, have caused many of us to judge our fellow men for what they believe. For the last 2150 years approximately, during the Piscean Age, we believed that God, a wrathful or gentle God was on our side against someone else. Despite various teachers who had taught the contrary, there are millions of people on this planet who still adhere to the old idea that their religion is better than someone else's. They believe their various religious leaders who coax them with words of love and manipulate them with fear, causing them to feel guilty. Many are angry with God and angry with themselves. Many feel that God has forgotten them, casting them out to be alone in this world. Still others expect God to give them whatever they ask for and curse God if nothing happens. Yet, in spite of all of this, there are times when each of us comes to a point of stillness, in which a still, small voice inside whispers, "…stop fighting; surrender and just be."

But, no matter how deeply that moment may strike a person emotionally, it is usually rationalized away within seconds by the conscious mind judging itself as insane, hearing strange voices, or losing control and becoming possessed by some evil entity. This erroneous thought causes millions of people to live in fear and doubt about their spiritual self. As a result, they cling even more to religious beliefs, swirling downward into the depth of despair, negativity and judgment. It is only at this point that they learn to surrender to their true self and to God.

It is interesting to note at this point, that the old biblical texts all speak of rules and regulations that should be followed in order to get God to be on a person's side. Should they fail in being clean, humble and sacrificial, then they were expected to be cast into hell, under the clutches of Belial (Demon). Still today there are religions that instill this belief into their congregations. Often those followers are full of fear that they will be cast into hell. I wish to state clearly here, that this is a man-made idea. There is not hell unless you personally create it here on Earth. Subsequently, most of the world has learned to live in a state of unrealized depression.

Now, in the Aquarian Age, the five new types of children will change the way we see religion. They will embrace the world of Metaphysics and Metaphysical/Spiritual training that comes with it. Their ideas about God are different. They each believe that they can listen to their own spirit, can talk to spirit entities, can connect with Angels and can be one with God. Each child, in his own way, will find a way to express newly found beliefs and to extend those beliefs into the minds of their children.

Religions are already transforming. Old Catholic ways have been changed on personal levels already. Though still considered a mortal sin to prevent pregnancy or to miss Sunday Mass or to not attend weekly confession, the Pope has no power over people and their decisions. Many Jews have learned to embrace Jesus and his message, while Muslims have learned to put down the sword and fight with words of peace. In the years to come, all will embrace one way of connecting to God. The Aquarian Age will bring in the full presence of the Second Spiritual Age, which is beginning now.

The Second Spiritual Age

While the first Spiritual Age was marked by innocence and ignorance, the Second Spiritual Age is an age of awareness. This Second Spiritual Age allows each of us to be consciously connected to all of creation. Our spirit selves will walk the Earth knowing how to make easy choices without fear and to live life to the fullest. We will still learn the hard way, but with awareness of why it is hard. We will not complain, but instead will embrace the lesson.

In this Second Spiritual Age, the Hero Children will work towards erasing religious dogma. They will search out false prophets and expose them through the media. They will also bring news of new philosophers and teachers. The Star Children will establish schools for learning Metaphysical subjects and will become teachers of various practices including integrating wholeness of health through healing and medical professions. The Indigo Children will express the arts and divinity in everything they create. They will be the philosophers and priests of the future, though their practices will vary as they evolve from the old religious ways. Their healing arts will be powerful and effective. The Crystal Children will establish higher levels of education in spiritual matters and will control the way these matters are shared and used. They will monitor the growth of spiritual progress, ensuring that historic records will benefit Mankind in the future. Meanwhile, the Liquid Crystal Children will become living, walking examples of sharing philosophy and practicing what they teach. They will be the public speakers, the organizers, instigators of new beliefs and the founders of new ways.

Though it might be difficult to imagine the role of this new generation in the new Spiritual Age, it can be conceived of by looking at our prior history in the First Spiritual Age. We have been living to survive and have struggled to succeed. Whatever the situation, Man has evolved somehow and, over countless times of being at the lowest point, has found a way to spring back into action to change themselves and the world. The Second Spiritual Age will be no different.

Each race has its own history and its own spiritual pattern of evolvement. Within each pattern, the five types of children will effectively cause a change in the world today. Each country still has a destiny to find its own spiritual level. They will all have to find a way to live, share and integrate by having different justifications for their existence. It will no longer be about obtaining land and property, but about obtaining balance in survival through trade, spiritual understanding and dependence with a view to sharing everything on this planet. As we look upon the wars, famine and cruelty that still exist today, that seems impossible to most of us. It should be noted that all of this 'bad stuff' is the only way in which to learn and to eventually surrender to something new that will lead to unity. In our destruction is our resurrection.

Each of these five types of children was prepared spiritually before birth for this leap in human understanding. There will be no doubt about their purpose and their spiritual journey in this life.

Spiritual Awareness

Though it virtually impossible to describe how every child will perceive themselves because they will all have their own *Soul Structures* (**See *The Rejection Syndrome***), it is possible to say how they will see themselves in general from a spiritual point of view.

The Hero Child will be able to meditate and focus on his/her body to internalize issues and to mentally heal by naturally controlling the flow of energy in the physical body, looking inward to see their spiritual lessons and to master their will. They will be interested in physical disciplines that allow them to tone and strengthen their muscles while keeping their internal organs active. Martial Arts are but one of many discipline areas they will use to understand the Tao.

The Star Child is tied to their Etheric Body and will constantly be aware of the ebb and flow of lessons that affect them emotionally and mentally throughout their daily activities. Their sensitivity to this body will help them discover more about the working of the human mind as it entwines with their spirit and the subsequent emotions that follow. In this way, they will face the illnesses of the world and try to correct them. They will struggle to transform fear and anxiety in an attempt to find inner joy on a conscious level.

The Indigo Child is tied to their Spirit Body. Everything they do is a question. They search for reasons for their existence and earnestly study their abilities. They are likely to have Out of Body (OB) experiences that validate their spiritual lessons. Their subsequent skills will be shared with the Hero and Star Children, who will naturally follow their lead. If the Indigo Child becomes ill, they will seek healing from Higher Powers while knowing that death is only the beginning and that life is for living to the fullest.

The Crystal Children are linked to both their Etheric and Spiritual Bodies Consequently they are in a battle between what they feel in an Earthly sense versus what their Spiritual Body makes them feel. This conflict between the two bodies creates an interesting twist in the way their energy flows. The conscious mind can be driven by left or right brain activities that can produce mental imbalances such as Bi-polar syndrome, but, given time and transformation, they will learn to harmonize the hemispheres of the brain and awaken a powerful ability to know how to bring wisdom from The Oneness into conscious thinking. In this way, they can tap into the Higher Mind of their true Spirit-Self to bring forth words of wisdom. The Hero, Star and Indigo Children will seek their guidance.

The Liquid Crystal child is attached to all five bodies: Physical, Etheric, Spirit, Higher Mind and Soul Bodies. In their own way, they will be the voice of all the other four types of children, as well as the voice of God within themselves. They may well be out of balance for many years to come, but will in time find unity within themselves that will point to a better level of existence where peace and harmony are the daily focus. They will be able to psychically channel their Spirit Guides and see the light and dark and know it for what it is; part of The Oneness (all that is created by God.) Their ability to control and transform energy will be extremely easy for them. They will be the healers of the world. They will eventually learn to easily integrate themselves with the other four groups of children who will respect them for their abilities.

These Five types will learn to integrate their lives though mental, emotional and spiritual awareness that comes from their spirit selves. They will not be guided by their Lower-self mental and emotional instabilities as they are today. As with all things, it will take time for these children to evolve into the new Second Spiritual Age. With the right parenting and good teaching, they will readily prepare themselves for things to come.

Though each group of children has a common growth factor, their abilities will interact and each group will develop abilities that cause a crossover in the way they show their talents and skills. For example, a Hero Child may develop latent psychic talents later in life, while a Crystal Liquid child may learn to appreciate their practical skills and develop leadership qualities that parallel the Hero Child.

There is no way of knowing how each of the children born in our future will develop their personalities and characters. But, you can be sure that for generations to come, they will be encoded with desires to reflect the mood and actions of God in form. Since God is believed to be 'The All Knowing," it should be easily acceptable to be open to discovering just what that is. In this Second Spiritual Age, the journey of such a discovery will be exciting for all.

As generations of all five groups of children come and go, so we will see a change in human stature. Their brains will process information at a much faster rate. Subsequently, our physical skills and mental capabilities will evolve. We have already seen Mankind literally grow before our eyes. In 1945, the average Japanese man was about 5 - 5.5 feet tall. Today, many are over 6 feet tall. The same can be said of other races. The more food we share, the more education we get, the greater the growth. The Aquarian Age will provide this for many of the Third World Powers who will take on much more authority than they have until now.

PART THREE

PARENTING THE RIGHT WAY

Chapter Nineteen

Communication

As explained earlier in this book, your children, in order to learn as quickly as possible, will be watching you very closely from the moment they feel with their psychic senses. They will sense your mood emotionally and mentally and will see aspects of you that you are unaware of as they study you visually as well as hear your tone of voice.

Young as a child may seem, remember that it is their wise old spirit-self that is doing the watching and listening, while directing themselves to copy everything you and others do. It does not take a genius to see the likeness in behavioral patterns in any one family. If it were possible to bring together all the members of one family going back over 100 years, though their individual ideas about the world and their personal points of view on things might be very different, you would still perceive the voice tones, body languages, facial expressions and habitual movements to be very similar, if not actually identical.

Because our genealogy and spiritual lessons will be in the mirror image or very similar as a result of our *Spiritual Soul Structure* coding, we can see a direct influence in the way we behave. It is not unusual for a grandmother to remark to her grand-daughter, just how like her own mother she is. Unfortunately, most of us put more emphasis on the negative likenesses. A child will sense many aspects of a variety of emotional responses coming at them from their parents and will in their own way try to emulate them. A drunken father's example may well lead to a drunken son. A yelling mother may well train her child to yell back. It is, therefore, very important that parents should be absolutely aware of themselves and the image they present.

Since a child seems not be able to understand all the problems and issues that his/her parent is dealing with, it is immediately believed that a child will not be affected by what parents say or do around them. This is not the case. For example, a couple was having a very intimate time with their two year old watching. A few days later, that child climbed on top of the mother and emulated the sex act, kissing her and jumping up and down. Of course, she was shocked and never repeated that act in front of her child again.

To a child, anything that is happening is something to copy. While they copy they learn something by having a personal experience that will mean something to them, even though it is not understood by adults. So, it is very important to spend time with your child, explaining what things really mean as they happen. Trying to cover up an event, hide truths, or simply deny an existence of some feeling will not go unnoticed.

It is also important to remember that your child is naturally psychic and will pick up your negative feeling of fear, pain and anger no matter how much you may be denying it to yourself. Being truthful to your self is exceedingly important. Building an honest relationship always leads to a great deal of love, trust and harmony that will ensure a true spiritual bond throughout life, no matter what occurs. A child should always feel that they can return home at anytime, not necessarily to be saved, but rather to find comfort and to receive emotional and mental support.

Children will always speak of some paranormal experience in the first years of their lives. They see 'Earthbound' Spirits – people who have become stuck between this world and the next, spirit animals, garden spirits such as fairies, gnomes and of course Spirit Guides, Angels etc. To them, these experiences are very real and very normal. They do not need to be told that they imagined it. This will immediately create separation with your child. Trying to understand what they see is essential.

Body Language

There are ways to truly communicate with your child and build a strong bond that will never break, no matter which type of child you have.

If your child has done something wrong, stand firmly in front of him/her. Do not tower over them, but rather squat down to their level. This way you are not overpowering.

Avoid waving your arms around as you demonstrate your frustration and anger, but rather point to the thing that is upsetting you and indicate to your child that you wish them to notice the object and the fact that it is (broken). By directing their face towards the item, you divert them from your wrath and help them focus on the problem.

Stand or sit still with your child while connecting mentally and emotionally with them through gentle touch. This will help them to develop an inner awareness that you are serious and that you have something important to show them. There are ways to truly communicate with your child and build a strong bond that will never break, no matter which type of child you have.

Always remember that their psychic impressions are running through their Earthly mind and heart. Try to imagine what they are processing by making eye contact in a serious but loving way. This will help you to calm down and feel your own inner-self and listen to your own spirit-self before you approach any issue verbally.

Be aware of the colors of your clothing. As you move, your child is making a direct communication with who you are by the type and color of your clothing. If you wear red and get really angry, they will associate you and that color as an indication of your mood. If you wear grey and look sad, they will always remember that in the years to come. These colors can affect the way your child relates to you on a day to day basis.

Your general appearance is important to your child. If your dress code is laid back, your child is likely to develop a lack of interest in hygiene and their own appearance as well as a lack of interest in organization or discipline. If you slouch and walk awkwardly, they will copy that pose and incorporate it into their own way of walking. Often genetic disorders seem to manifest later in life because of the child's original observation of mother, father, siblings and peers. It should be noted, that many children do not use their feet correctly. Toes can turn in and balance is often placed on the heel and not on the toes when walking. A parent can help a child to walk correctly in early dance classes. Bad posture follows in later years if this is not corrected.

The way you eat food can also be an indication of body language. Your child will watch you and copy you. If you eat slowly, you will digest your food well; if you binge, your body will become obese. You child will copy these habits and will develop accordingly. If you are anorexic, your child may well eat for you and develop a great hunger that cannot be satisfied. All these eating habits are tied into emotional-psychometric feelings that your child senses in the first years of life. The way you sit and eat; how often you get up and down from the table will all be copied. Often a mother will get up from the table many times to serve various foods to other members of the family. The result is a child, who will not stay seated at the table; who constantly cries to get out of the highchair. By placing everything on the table before sitting to eat, and remaining seated until everything is finished, your child will learn from the comfort of your body, that being still is a good way to be. This lesson is learned during breast feeding, but is soon forgotten, once action in life takes place. So, reaffirm this lesson of being still to eat.

Since we all have to communicate with others in front of our child, it is very easy to be aware that a child is watching your every move. If you are in the midst of an argument with your husband or family member, your body language will give everything you are feeling away. I once saw a man holding his crotch while yelling abuse. That speaks loudly of a lack of security. We often see young children holding on to their private parts while watching other children at play. It is important to be aware of how you present yourself in front of your child when dealing with other people. If there is discourse, then your child should be reassured afterwards.

Walking away from a child during their time of need, even if it is a tempertantrum, will affect your child's sense of self and set them up with an idea that anger means separation. If you need to create space from your child, simply talk quietly, kiss them and then say "I will be back in a minute." This gives you both time to calm down and allows your child to learn that walking away is not a final separation. When you return, you show your delight in returning with a good steady walk into the room and a happy face smile, even if your child is still crying. Simply say something in a calm way, such as "Wow! You are having a good cry, aren't you?" Then arrange yourself in a comfortable position and continue to dialogue quietly.

Since it is impossible to catch all your moods and all your subsequent body language behavior; obviously it is better to try and develop a calm nature within yourself. Find a way to solve issues with easy solutions that do not allow you to ponder, muse or reflect over negative emotions. Take a more philosophical point of view and say to yourself, "This is a hard lesson, but I/we will grow as a result of it and 'where there is a will, there is a way' to solve this problem. I am open to change and will adapt as needs be. I let nothing set me back. There is only room to grow and move on". Such a positive attitude will develop good body language. You will be amazed how other adults will acknowledge the changes in you and how your body seems to be more fit.

A good sense of well-being generates positive energy. Your child will emulate you. If you are weak, your child may develop a shallow breathing habit that could result in allergies. A child watches the way you breathe and the way you use your breath to speak. If you generate very little power in the way you breathe in and out as you speak, then your voice will be quiet. If you are constantly inhaling at a fast pace, then you will have a loud sharp voice. Ideally, a good use of the diaphragm will generate a healthy breathing pattern that your child will copy.

Perhaps you have a weakness in the eyes, blink a lot or wear glasses. Your child will associate the looks you give him/her with the way you use your eyes. If your eyes dart around and look harsh, you child will look away; even cry for no apparent reason. If your eyes look fearful, your child will become nervous and want to walk or crawl away. It is said that eyes are the 'mirror of the Soul', and your baby is watching your 'Soul' for every little thing you feel. To avoid creating insecurities in your child, take time to develop eye to eye contact during times you discuss something nice. Later, when they learn rights and wrongs of life, they will not shy from your gaze, but rather stare back at you looking for the real truth. So, be sure you are coming from a truthful position and not just a nagging one.

Since we do not flap our ear like dogs or cats, it would seem that our children are not watching them. But, the opposite it true if they can see them. When you are angry, the pigment in your ears becomes redder, just like your face. When you blush, your ears take on a pink hue too. Since your entire skin is an organ and it responds to your emotional state, your child is watching that too. If you are afraid and turn white, your skin gives you away.

Perhaps you have emotional habits, like running your fingers through your hair, or biting your nails. Your child will once again copy you or develop another nervous habit to emulate you. Since you do not want your child to copy this type of behavior, it is best to heal yourself of these habits. Likewise, if you are a drug taker, whether pharmaceutical or from the street, your child will watch as the effect of the drug changes your skin tones, your breathing patterns etc. In the years to come, they may well develop the same needs. Those who are taking suppressant drugs are often not aware of the changes in body language that occur. Your child is! Clean up your needs for these drugs and your child will learn to communicate honestly.

Since we are talking about body language and habits you have taken for granted, many of which you have learned from your own parents, siblings, grandparents and babysitters, it is important to try and educate all your older children and family members to be equally aware of how they behave in front of your small child. If you have been through tremendous trauma during the birth of other children or have been divorced etc., then your new child will be spiritually aware of this, but in physical ways, they will respond to the other children in the house who may have become completely unbalanced as a result of emotional trauma. The elder siblings will act according to their own perceptions of how they were treated and their own personal observations of you and their family at that time. Since it is not easy to help them change their point of view, it should be explained to them that their new sibling needs them to find a new way to integrate their awareness with him/her. Encourage them to find their own way to make eye contact and take responsibility for helping in the care and attention of this small child. It is often amazing how unstable children suddenly become little responsible adults when a new baby arrives.

If the older siblings become jealous of all the attention that is given to their young sister/brother, take time out to be completely alone with them and to apply the same techniques as listed before. Their inner small child is still looking, watching your every move, listening to all the variety of tones you use and the way you speak to them. Reach inside yourself and let that instinctive mother out. Give them your spiritual essence and reach their spirit to find a common bond that will create mutual love, respect and a spiritual bond.

When your young child is looking at you, they are also looking at the colors of your aura. If you are thinking heavily, dark yellow will emanate from your Crown Chakra (Vortex of energy around your head/halo). If you are calculating, a variety of blues will emanate from the left side of your Crown Chakra. If you are teaching your child; those blue and yellow colors will flow across your head and out of the right side of your Crown Chakra. Whatever, you are thinking, whether distant past, immediate past, present or future, those thoughts are emanating a variety of colors across the top of your head. Likewise all the emotions you recall as you think will also flow out of your Crown Chakra. Your child is watching a color show that can be likened to a firework display. As all this happens, the remainder of your body is doing the same for each of the other Major Chakras (*see my books on* **Crystal Acupuncture & Teragram Therapy and Psychic Development**). Your own child is immediately processing all these colors with your moods, your words, your appearance, the sounds you make as you walk, sit or stand etc.

They are a spiritual detective, who is awakened to all the coding of your *Soul Structure* that was absorbed by them while they were inside you. Believe it or not, even when you are distressed and upset, there is plenty of comfort in discomfort for them. This is why many children grow up often feeling that family unity evolves out of anger and other emotional disturbances that make a mental equation in their mind as equaling deep love relationships. They later grow up to create relationships that are built upon insecurities and self-destructive ideas.

Try to solve any insecurity that you may have as quickly as possible by finding solid solutions at the same time, alleviate your child's insecurities with simple, but sound dialogue that provides assurance. Giving toys or candy to calm them down is not a good way to settle their issue. This leads to a great need for 'things' to replace love in their future, when all items are seen as a symbol of love. Often, adults who require constant attention are habitual shoppers, who grow angry with every purchase as they realize repeatedly that they cannot buy happiness. Any item that you give to a very young child to appease their mood will be read as meaning much more than you may ever know. Its shape, feel, size and color all have some form of meaning to your child. Everything you give him/her is an extension of you.

Talk, Music And Song

Once your child has watched you and copied your physical movement, he/she will copy the sounds you make. It seems to adults that talking is easy but in truth, it is very awkward to form the lips into shapes, curl the tongue up and force air through the larynx while controlling the shape of the mouth. A baby will quite naturally make simple one syllable sounds, such as Da, Ma, Ta, Mi, Bi, He etc. These words are formed as though singing the scale, Do, Re, Me, Fa, etc. The more complicated sounds like Ess, She, Bla, Qui, Thr, etc. come much later once confidence in moving the mouth and forming the words is understood. Sometimes young two year olds talk 'Double Dutch' rubbish words, but in their mind, they are talking. Adults may laugh at this, but seriously, the child wants you to appreciate that they are communicating with you. Giving simple responses, such as, "Wow!" "Great!" "Good," etc. will help a small child and encourage them to say more. Every mother does need to spend time helping her child to form words. Basic elocution lessons are important. I remember when my eldest child found it impossible to say "Thing". Everything was pronounced "Fing", and still today the word "Somefing" flows out when he is tired. This verbal slip validates the fact that we all go back to our first awareness of speech and will use those words often without even knowing we are doing it.

Some children love to sing first. Their little voices lilt as they dance around delighted with the sound of music while emulating those sounds. Later when they learn to make words, their voices sound musical. A little later, they learn to sing nursery rhymes very early and often express their emotions with high pitched wines and low crying sobs. Even after many years of adulthood, those early sounds will still attract them and give them comfort, despite the fact that they have consciously forgotten those years. When the elderly develop Dementia or Alzheimer's disease, often those baby year's experiences arise when listening to music and when encouraged to sing those baby nursery rhymes, they remember them all. It can also be heard when an adult is whimpering. The Voice goes higher and sounds like a child whining.

Everyone can sing. Our speech pattern is based on melody. We speak in rhythm and tones which resonate and reverberate throughout our body as well as across the room. No matter what the language, those musical notes we make are recognized as examples of our personality and character. So, adults who insist they cannot sing are in fact lying to themselves, as speech is a form of song. Every song we love to hear was originally a poem which was later put to music that embellishes speech into a melodious communication. Everyone listens to their favorite songs, which seem to always have a special message of comfort for them.

So, the formation of words, the way those words are used in structured sentences is important. Any sentence that is supportive, encouraging, educational and informative is highly important to a child. So long as it is given with good tone, resonation and loving, nurturing vibration, your child will listen and learn. If the sentence is sharp, or flat, just like when music it is out of sync or is in discord and uncomfortable to hear, your child will shut you out and ignore what it is you are trying to tell them. So, if you are angry or sad, it is important to make your voice light and breezy so that communication sounds melodious and appealing. Then your child will listen to you. You can then explain that you are tired and have a problem. Knowing this, they will be better behaved. Use simple words that are easy to understand. Don't spill out all your issues at once. Give them one at a time, allowing your child to assimilate and understand you. If there are many issues, it may take several days for them to be sorted out.

Children between the ages of 5 months and 5 years are not able to consciously remember and associate all the sounds they hear until their critical mind is developed around 8 years old. Until then, the critical mind is simply a momentary thing that easily helps a child make up his/her mind for the moment. "I like it today," or "I did like it, but now I do not," One week later, they will have a different answer or have simply forgotten that they had an opinion. Yet, despite this growing consciousness in the brain, their Spirit-self is recording everything, just like a tape recorder and those early sounds and little things that happen are never forgotten; What is stored in the subconscious will have an influence on them in the way they develop their personality and character as the years pass.

Teaching your child to make sounds and develop speech should be fun. Never make fun of them if they are having a hard time pronouncing something; it will affect them for the rest of their life. Any negative association with sounds learned, even when they finally master pronunciation, will become key words in dialogue as the years pass that will make them feel insecure. I, in my own life, often tried to say the words "able to" which came out as "abbleto". Even though I knew I could break the words down and say it slowly, whenever I spoke fast those two words always came out as one, sounding more like 'appleto', so with much practice, I mastered those words, but even today, if they come up, I remember my deep-seated sense of insecurity around those words. My subtext thought was always "I am not able to...."

This type of thinking is very general with everyone. We all have insecurities about the way we phrase and enunciate, even though we never think about it, unless put into a threatening situation.

To aid your child in developing speech, encourage singing, lots of music and plenty of support in repetition where praise is given along with the statement "You are so good! You can do anything you try to do!" This simple statement, when repeated many times will hypnotically teach your child to believe that they can do anything they try to do and will do very well in life.

Asking your child to sing along with you, will help them to resonate and speak loudly too. However, you do not want them to be shouting their heads off, deafening everyone, so singing softly and moderately and then loudly will help them to modulate their speaking voice.

When I was a child, I was told that "Children should be seen and not heard," and to "only speak when spoken to." Since the adults were always speaking loudly and abruptly over one another, I did not develop my small speaking voice. I was often told to "Speak-up" as a result of being afraid to over-step my mark. I did not know when it was a good time to speak, and when I did speak, if I was saying the right thing! For most of my youth, I was considered a quiet person, who did not speak up when approached by people of authority. By the time I was twenty-five, I had figured it out that I was finally an adult and had plenty to say. I surely let rip then. This was not a good way to learn to speak up.

So, give your child plenty of time to speak with you. Everything they say is important to them. Involve them in conversations on their level. Lead them with questions and give them answers when they ask for them. Make sure those answers are clear. "No, you cannot have ice cream because you will be eating lunch soon, but when you get hungry after lunch we may be able to buy some later." This gives your child an understanding and hope that what they want can be sorted later. Hope is a very important part of development. Through the years, your child will hope for many things, which will motivate them to try much harder to obtain their goal.

Clapping hands, waving hands, jumping about, doing exercises and learning to control their bodily functions is all tied into speech. Bladder control is one of the most important issues and every child has to learn to control themselves and wait for an available toilet. The moans, the movements and the sound of running water have an effect on the brain and nervous system. Dialoguing with hands supporting expression is also an important part of communication. We are all taught to point out something as we see it, but later are often told it is rude to point. These inconsistent associations with words, feelings and bodily functions can cause an enormous amount of insecurity. Once again, meaningful dialogue is important where manners, tact etc. are explained to a child. There can be many ideas floating around in a child's brain, causing them to assume incorrectly, resulting in wrong beliefs that later in life, create many unnecessary emotional crises. It is not unusual for adults to develop weak bladders during insecure times.

It is hard to know how your child will respond to your words and phrases. As a parent, you need to listen to your intuition and then follow it to the best of your ability. If you learn this, you will be amazed how much information will pour out of yourself. You will not only communicate well with your child, but also with your peers and family members

Throw a party; then allow your child to integrate with the family members and friends. Sing and dance and make sure your child joins in. These early exchanges will ensure that you develop a child who is positive and extroverted.

High pitched screeches are normal sounds for very young children. They express their delight with such sharp shrills, that it can be deafening. These high pitches are the early sign of a soprano or tenor singer. Ensure that schooling around singing is encouraged.

Continuous crying is a sound that should be discouraged, as it causes a child to self-hypnotize fear around the sound of those notes as they listen to their own voice. Crying and talking at the same time turns into conversations that are associated with whining and later still, with a negative attitude during dialogue. Since the brain records every sound, emotion and thought, your child can listen to his own self-talk and create negative beliefs that will be hypnotically accepted.

These early childhood ideas can ruin a life. I have researched this fact over the many years I have been practicing healing and as a result created a hypnosis CD called **You & Kandoo**. This is a channeled work that has two tracks, one for girls and one for boys of any age. Adults love it, as it reconnects them in positive ways to their inner child who is still crying out for some positive way to exist. Younger children become A students, while 'abnormal' children improve their skills at learning. This CD is recorded in the style of a fairy talking to a young child. All children believe in fairies!

Since every child is in a state of hypnosis for the first five years, they will be influenced by statements you or others make such as "You don't understand," "It is not like it appears," "Things are worse than they appear." When a child listens to his peers, he/she will copy them to try and assimilate their feelings so that they can say they have a feeling of belonging to the group. That group could be family and kindergarten. It is not unusual for young children to copy sayings they have heard from others without knowing the meaning of the sentence. Swear words are often bantered about, having picked them up from their own parents. Those swear words have a lot of power and energy behind them when said, even if whispered under the breath. Of course that is a No! No! But chastisement in the use of that word can be totally misunderstood if you keep saying it too!

The sounds of breath are also full of meaning. Your child will listen to the way you breathe and interpret those sounds to mean something from their point of view. They will copy those breathing patterns throughout their life, even though they may do many things in different ways from you. I once had a child who was learning to dance in my class. She held her breath every time she moved. When I asked her why she did that, she shrugged her shoulders and said, "My sister does it too." It was then that I learned that families breathe in similar ways; some shallowly through the neck, while other breathe from the mid lungs; a very few have learned to breathe deeply from the diaphragm.

When listening to a child talking, watch to see how they breathe. If they seemed to talk a mile a minute without stopping to take a breath, then you can know that they are afraid they will not be listened to. If they speak very softly, then you can tell that they have a fear of rejection. If they shout and emphasize one word, that will be a sign of impatience. Still others will correctly breathe, pausing between sentences to ensure you have listened.

If you do not pay attention, they will tug on you; stare at you uncomfortably until you give them your undivided attention. They will then, speak clearly and precisely. This is a sign of a strong willed child. These individual ways of breathing and communicating are all stimulated by the child's *Soul Structure*. You can have several siblings with different ways of using their breath when speaking in one family or alternatively find them all breathing in the same way.

Another aspect of speaking is moaning and groaning when a child hurts him/herself. Various sounds such as: "Ouch! Boohoo! Sniveling etc. can also set a psychological pattern within your child's mind. Over the years whenever he/she hurts him/herself, there will be a strong association with those sounds either inwardly or outwardly. The moment any kind of pain is felt, a state of hypnosis is created. In an alpha state your child can tell him/herself that life is 'unfair' or something similar. Often, anger or self-pity is expressed with the same tone as an "ouch or boohoo." That kind of self-talk will lead to a life of suffering if it is not corrected with love and support.

So, it is important to realize that every sound we hear has an emotional tie into some kind of joy or suffering, which will be expressed outwardly by a child as they grow. Those sounds are tied into beliefs and most of those beliefs can be so rigid as to cause destruction, self-depredation and martyrdom.

It can be hard on your ears to listen to your child's squeals of delight, but it should be noted that those simple pleasures and the sounds your child makes will burn a deep memory of well-being and confidence into their brain. So, in such moments of great excitement, try to teach your child to use their other senses of vision and physical sensations to help them become quieter and to express self with some modicum of control.

Every child likes to listen to the sound of their own voice. They want absolute attention at all times, and will use any kind of communication to get it. They may shout, scream, speak tenderly or simply sing to make you notice them. Your emotional response and the way you dialogue with them physically will lay the foundation stones for them to integrate with others. As they mature, the one voice they will always hear first is their own. So, it is important that they learn to speak to self in positive ways.

Since every family is unique. Different words have different connotations. When two separate families are united though marriage, there is often a misunderstanding between families, bride and groom, as dialogue becomes serious. Each is listening to the same words, but hearing different tones and resonations and interpreting those sounds according to their childhood memories and upbringing. Often arguments result. If you change your own way of dialoging and integrate your partner's pattern of communication, then the sounds and tones will take on a different meaning. Then you can truly listen to one another. Your family history is only as important as the lessons you have learned, otherwise, it has no importance.

The Hero Child will pay attention to success stories; The Indigo Child will pay attention of fairy stories with a moral; The Star Child will love stories about your personal experiences in strange circumstances. The Crystal Child will require a very small short story relative to the moment. The Liquid Child will listen to any story and then tell you his/her version. Whatever the story, their imagination is working over-time.

Touch

There are two types of mothers: the touchy-feely informal, hugging mom or the distant, mental, always organized mother. That is not to say that they do not intermix, but rather to say the one aspect is more dominant than the other.

If mother insists on having an organized life with routines, discipline and structures that she considers a priority, then her child will learn to play along. He/she will never really understand the logic of his/her mother's ideas and routines and may well resist often, but ultimately will obey. Later in life, her children will be more distant and less likely to want to be close to her.

The mother who loves to hug, pamper to her child's every need, will often have little or no structure to her day, unless she is working. Her child will learn that she can be manipulated according to her child's wishes. As he/she grows up, mother will be over-ridden time and time again. She is likely to feel as though she has sacrificed her life for her child.

Both of these aspects of motherhood need to be integrated. A child needs to know the power of tenderness and compassion, as well as the comfort and control of closeness. But he/she also need separation and discipline too.

During the first hours of life, your child was held by you and learned that this was a truly comforting and simple way to be with another human being. As he/she has grown, autonomy has created a separation between you both. However, that separation should not be an excuse to never hug.

Over the years I have met many adults who are unable to hug. When held, they pull their body away as soon as is possible. They pat gingerly on the back of the hugger while saying something simple to make an excuse to break away, such as "I must go now."

I have met other adults who hang on to me for dear life, lest I should abandon them before they are ready to be released. Their hugs are so tight, that I feel uncomfortable and made to feel responsible for their happiness.

The ideal hug is a moment of inner peace, when two people touch, hold and share energy without any thoughts or purpose beyond just being connected. When you first held you child, there was no thought, only a need to hold, touch and share energy. The same can be said of holding a new baby animal. It is in our nature to touch and to be touched.

To teach your child the importance of tender touch is invaluable. Priceless! Through the years, they will explore their emotions and reach out to so many strangers with a welcoming hand-shake or hug that will make them feel at ease. In this way, you child will become an extrovert. Those who are not touched, often become introverted.

Some adults are unable to touch unless there is anger. A slap on the thigh, a punch on the nose or worse is also an extreme lesson in touch for any child. Some children learn to pinch, jab, pummel and fight, while others stand by and watch. Once an adult becomes involved, the aggressive child may be touched/manhandled/slapped and in this way he/she gets attention, even if it is just momentary. By the time they are adult, they will not think twice about hitting their own children, passing on their understanding that emotional pain and anger equals love.

Everyone loves to tickle a child. Giggles are wonderful and joyful for a while. When the nerves in the body become oversensitive, then tickling becomes a discomfort. It is important for an adult to know when a child has had enough. If too much tickling happens, a child will develop barriers against being touched anywhere. This can have a bad influence on sexual experiences later in life.

Sensory Touch Therapy (STT) was created by my husband and I. It was easy to see how a very light touch could both tickle and cause extreme sensations of pleasure. When you first held your child, caressing his/her limbs, those extreme sensations were being recorded with his/her brain. As the years pass, if this kind of touch is denied, they will crave it from other sources. Because of sexual growth, this is often mistaken for a need to have sex at a very early age. STT is a therapy that any parent can do to the arms, neck, legs, hand and feet without causing any emotional discomfort. The more laughter and joy your child gets, the more confident they will become. This soft touch can be varied with light scratching which can be equally pleasurable.

If a child is denied physical touch of any kind, they will grow introverted and eventually isolate him/her self from society. Often these children become ardent students who drown their sorrows in text books, bedroom activities such as computers or loud music.

A child with a balanced sense of touch will have a schedule that involves sports, dance, the arts, music, clubs, cub-scouts etc. They will learn to quickly interact and develop leadership qualities.

In the old days, before people cried "molester" when a teacher held a child, many a lonely child had been comforted by a teacher's hand. Now today, this modern society has destroyed much of the closeness that a child needs in which to be trained to survive. It is now doubly important that a parent hold, nurture and love their child with a good balance of discipline in action.

When your baby was born, you held him/her close without any sense of fear, embarrassment or self-consciousness of the opinions of others. This true sense of physical contact is the top priority for any new-born. The way you hold, caress and nurture you child will formulate a great desire to be held and love for the rest of your child's life. If you suddenly cut off that contact, then your child will seek physical contact elsewhere.

Since most mothers have more than one child, the eldest child usually feels left out as he/she watches attention shift from self to the other siblings. In order to be a part of what is occurring and to be touched, this child will either become a 'little mum or dad' fixer child or develop a dominant aggressive nature which can result in yelling, hitting and pushing the younger siblings. It is important to take time out to sit with each child and give that hug, comb hair, take an interest in what they do and wear. Smoothing clothes, measuring shoes to see if they fit, trying on coats etc. all add up to physical contact with another person, especially mum.

By the time a child is a teenager; physical contact has taken on a new meaning. Motherly love is still needed, but in different ways. I often asked my children for a hug, pretending that I needed one. This gave them an opportunity to have that physical contact, without being childish, while at the same time giving them an awareness that touch is important. All my sons are very physical and believe very much in hugging their own children. It is the same for everyone; we copy our parent's ways as we raise our own children.

I have seen many people over the years who complain about not having had enough contact. They want to hug, but cannot for fear that they will be rejected. It is wonderful when you can get all the family into therapy and address this issue. By the time they have all overcome their inability to hug and enjoy the close contact, all their anger and fear is washed away. Hugging brings out the child in all of us.

If there is no physical touch given to a child from 2 years on, other than aggressive movements, that child will withdraw from society, become mentally disturbed and incapable of assimilating various experiences. Unfortunately, we have seen a lot of children in various parts of the world who are without love and hugs. It is time that we corrected this for the sake of our future.

Hero Children will seek physical contact by becoming the fixer child and can easily become frustrated, yelling and slapping others around. They need a lot of eye contact, hugs and support during their adolescence. Later, they tend to give more hugs, rather than to receive them. This is their way of being close and receiving physical touch that satisfies their craving for love.

Star Children will demonstrate their talents to get attention. When they do well, they expect praise to be given in melodious tones, along with a big hug which may not last very long. Those momentary hugs are all they need until later, when more praise and support is needed.

The Indigo Children will ensure contact by making sure that they have a seat next to you and can engage you in conversation. They will stroke you, cuddle up close and climb all over you to ensure physical contact. As they grow older, they will fight for a space to be close to someone they care about. Physical touch is vital to their well-being.

The Crystal Children do not need so much physical contact. They are aggressive by nature and will demand hugs on their terms or when they hurt themselves. They accept plenty of contact in physical activities such as sports, dance and creative pursuits. From a parent's point of view, hugs seem to be a very occasional thing, however, to the child, that one hug will last for days, giving them a sense of belonging which provides and anchor for them.

The Liquid Crystal Children are constantly in need of touch. They will whine/scream to get attention if they feel neglected. Some will become people pleasers to gain that hug. Whatever they do in the future, it will always stem from physical contact. The more positive the contact is, then the greater their awareness of their spiritual self will be. Their perception of touch will formulate skills that will be used in their working life.

Outside Sounds

We are all aware of the sounds we make as soon as we learn to open our mouths and cry or gurgle. But, what of the sounds that come into our brains? How do we discern those sounds? What makes certain sounds comforting versus fearful? Some say it is the sharpness and suddenness of unexpected sounds that disturb us from our placid moments, while others say it is the sounds of people who scream and yell. Over the years, I have observed myself, my children, grandchildren and other people and have come to the conclusion that it is not the sound, but rather the vibration of that sound that always creates a shift in the energy of the Aura. That shift is then received by the brain, modulated and stored as interference or a reminder of something that has been heard and felt before.

Unexpected noises, such as squeaking or slamming doors can make a child jump during slumber. However, we are unable to ask a sleeping baby if they are now permanently afraid of those noises. I remember how my children responded to the irritating sounds of police cars as they flashed by with their sirens. They simply ignored it during the first two months, but when in their second year, did stir or jump while sleeping. Still later, they considered those very same sounds to be exciting and something to watch and listen to. It was only after they had learned what the noises meant that they developed a fear of police, should they break the law.

We all learn to associate outside noises as part of our background disturbances that can be recognized and explained away. If we choose to pay attention to noise, we become angry and frustrated. For example, having someone outside your house digging the road up all day will cause you to become tired and restless. If on the other hand, you sit with your window open and listen to soft music waffling from your neighbor's house, you may feel light and happy. Low vibrating noises can cause one to fall asleep as does the rhythm of a beating heart or the sound of easy breathing.

Since the conscious mind is very selective and our emotions are attached to those selected mindsets, we can choose to be disturbed or not. Much of what we learn about sounds processed by the conscious mind is stored away in the subconscious mind awaiting some future consideration. For example, if you hear the ice cream van playing music and then later in years hear a similar sound coming from your radio, you may integrate those two sounds and then stimulate your consciousness to make a choice of ignoring it, or suddenly deciding to go for some ice cream. While making your choice, you will activate an emotion that requires it to be noticed, sorted and settled. In this way, we begin to understand our associations with sound.

Because there is not a passing day without sounds penetrating us, our minds are constantly shifting and sorting out our reactions to them. If we have had a series of negative events related to sounds, then we are likely to be emotionally negative and depressed. Most adults focus on arguments and rationalize that they are upset through having had to experience such an event. The truth is, it is not the argument, but the tones and resonations of the sounds that disturb. Watching movies is one way to get upset or feel happy. By being in observation, one can isolate the way one responds to sounds. Many of the moods you develop are rising as a direct result of a sound you have heard. You do not need to wait for a headache to know this. If you watch your whole body, you will feel an energy shift long before pains and emotional discomfort arises.

Items like drills, electrical equipment with high pitched wines, radio waves and telephone signals are constantly bombarding our bodies. Even if we do not hear them physically, they are affecting us. In this modern day, no one is free from being attacked by some kind of sound. Even the deaf are disturbed by them as they feel their auras shift and adapt with the use of highly developed Psychometry.

But do not despair. Your baby's encoded *Soul Structure* has been prepared for this. Each child born today is ready and able to handle all those loud noises and to categorize them into two groups: important and unimportant.

Children today are born with technology as a way of life and with it an ability in the brain to deal with all this new information. Their brains are turned on within hours of birth. They are listening and vibrating with every sound. One of my grandchildren was only a few weeks old. We had to go to a family wedding and, as I held my grandson, I felt his body vibrating like a drum. I wanted to take him away from the loud celebration music and give him some peace, but since he seemed to be sleeping through it all and his parents were unconcerned, I decided to make light of it. Today, that child likes to have lights on, sounds around him and plenty of action. He is far from insecure, and does have a very inquiring mind at the grand age of three.

So, if you are of the older generation and find it hard to concentrate when loud noises are going on, take a leaf out of the new generation's ways and learn to switch off or associate that sound with some other thing that is more pleasant. Learning to be single minded is a way to rest your brain, which in turn will prevent you from being so tired. The ultimate test is to be in one of the noisiest places possible, and be able to focus on just one sound that you can hear clearly. This is what our children can do.

Hero Children like to have their music up loud, work on the computer and have a conversation at the same time. During all this activity they do not miss a beat. They are busy preparing for something else that seems to be even more important. Homework gets done with the music blaring as do chores and research on websites, too many to mention.

Star Children like to focus on one sound at a time. While there may be plenty of children making lots of noise around them, they will cut them out and focus very easily on what they are doing. They can be extremely single minded and can be known to complain about sounds if they are too loud or sudden. Often they will jump when approached unexpectedly. They have a love of melodious sounds that seem to inspire their minds to do great work.

Indigo Children are extremely sensitive to sounds and are likely to have built up a library of references within their brain. Many sounds will have been categorized into groups of 'sounds like' or 'remember when.' They have a strong emotional association with all sounds and can be particularly disturbed by strange and unusual sounds which may cause them to become afraid and in need of comfort. Their psychic gift of clairaudience is abnormally high, often creating a fear of seeing the unknown. These children love the sounds of nature.

Crystal Children are the noise makers. They love to listen to as many noises as possible: they like drums, musical instruments of any kind, whistling and the sounds of engines and motors as well as the noises other children make. There is not too much that disturbs them unless they have been educated to be afraid of certain sounds. In their world their belief is, 'what comes at me, I can take it and send it back louder than before.'

The Liquid Crystal Children are still being discovered. Their innate abilities to listen to sounds that are beyond human range are paranormal. They can hear the voices of angels, Spirit Guides or whispers in the other room. At the same time, they have the ability to hear a sound(s) once and repeat it correctly. Music is an art they master easily as is singing, or any other noise that can be made with the hands. Later in life, they will be the engineers who can listen to an engine and tell what is wrong with it without using technology to find out. They have an uncanny way of knowing when the sounds of an aura are changing as an indication that a person is about to become ill; equal to that of an animal. Their senses are acute and sound will always play an important part in their growth. They can be easily deafened and hurt by emotional tirades or loud unexpected noises.

Every child will listen to the sound of their own voice first and will lay great inner responsibility on what they tell self. Sometimes this inner running dialogue can be harmful. It is best to get your child to communicate their thoughts. It is also possible for a child to listen to the voices of Earthbound Spirits and Spirit Guides in their heads. So, do not discount what they say they hear. Discussions will reveal a great deal.

Visions

What these five groups of children see is not necessarily what you and your friends see. There is an innate ability to observe life with a different dynamic. For example: you may see an old lady crossing the road, struggling with her basket as she hobbles along. Your child may see a woman who has been troubled all her life and who is bent up with hatred and pain.

These children even have dreams that are different. Their nightmares are not necessarily of this world. They can be dreaming about dark monsters, far off planets or strange people that live and look different from us. When they try to tell you, your imagination will run wild and you will think them imaginative. But are they? What else is in the Universe that we do not know about? All these children are preparing for a future when we not only learn to live with one another, but also learn to live with other species.

Throughout time, we have speculated about our uniqueness and the possibility of another species on another planet. Our children of the future know more than we can ever understand. They know where we have been, what we have done and will apply that innate wisdom to what is to come. One day Star Trek will be a reality. Everything that Mankind can imagine will come to pass.

These five groups of children are encoded with the influence of the distant planets to have a different focus on life. As Pluto, Uranus, Jupiter and Saturn move across our skies, they will contemplate a new world. A world where all are fed, clothed and employed. Today, as I write this book, the world is in global recession and war. Our lessons are as plain as the noses on our faces. We must look beyond the nose, into the eye of God. We must look beyond war and materialism into the Soul of Mankind. As long as we compete and destroy, we annihilate ourselves. Our destructive twentieth century has been a long and hard lesson. Now in the twenty-first century, we must learn to share. Our children will define the future. Their children will bring a better world into form and their children will begin to put what has been learned into practice. All this begins with vision.

The Hero Child will stare at anything that moves to understand its existence, purpose and use. If something cannot be seen to be of use, then it will be destroyed, transformed or turned into some other kind of use by this child.

The Star Child will focus on how things were done and then examine what is current to see if there has been growth. If there is none, he/she will destroy what is in existence, making sure first that there is something to replace it. Their vision of how to make these changes will be fixed and exact.

The Indigo Child will have a much lighter approach. They will look at Mankind spiritually, and observe the decline of spirituality. Their motivation for change will stimulate them to educate the innocent and to direct the leaders of this world by being their conscience. They will describe their visions and inspire others to see for themselves.

The Crystal Child sees everything they touch as something to experiment with. They will look at the economy and structures of the world with a mind to destroy systems that are ineffective. They will often have a narrow vision of the changes they make, but with guidance from the other four groups, they will be the builders of their dreams into reality.

The Liquid Crystal Child will have a far reaching view of the world. They will make plans and plot to create something that is influential to the other four groups. Because their visions are so radical and unusual, it will be hard for them to convince the others that their way is best. Given a few centuries, their visions of a better world will come to pass.

All these children of our future are of the essence of God in a new form. Their abilities and skills will develop them physically, emotionally and mentally. They will grow taller, while their brain capacity for learning will expand and, as it does, their spiritual essence will manifest in emotions in such a way as to stimulate them to work towards a world of peace.

Many children can see images in their head. Their psychic sense of Clairvoyance is often amazingly sharp. Accept that you child can describe pictures to you and so always give credit to what they see in their mind. You may learn a thing or two from them. Often their images carry a message in a symbolical sense.

PART FOUR

GENERAL INSIGHTS IN BRINGING UP BABY

Chapter Twenty

Creative Talents

When I stop to ask my students In Japan about their talents, most will say "I have none. In England, they respond with "I have no time!" In Europe they inform me that "No one appreciates the talent or the effort, so why bother?" While in America everyone will say, "Oh, yes I have a talent/skill and I use them now and then when I can fit the time in." So, I find myself wondering why everyone believes him/herself to be without talent or skills or to feel they do not need to use them when they do have them.

The answer lies in creative education. In kindergarten, all children are exposed to their create skills. They learn to cut, stick, draw, mould, color, model, but by the time they enter their first grade school, most of what they have done is pushed aside in lieu of learning to read and write. Often the left hemisphere of the brain is forced to override the right side with the end result that a child believes that he/she is incapable of doing any creative project really well.

Since birth, our individual visual observations are ingeniously stored and calculated to mean something emotionally and physically. Those visions are correlated with thousands of other images that are later acquired providing an individual perception of a variety of meanings to one's personal life's experience. To some children what they see is automatically something to explore and in so doing, immediately develop a tactile desire to do more, while others prefer to observe and guess what the outcome might be. Still others wait until someone leads the way as a model of demonstration, washing away fears of inadequacies. Even further out on a limb, are the non-believers who doubt the very existence of truth. They wait patiently on the side, looking for proof in a never ending loop of hope. Whatever a child's idea is, it cannot be emotionally understood, unless they see and experience something physically. No child is alone on an island. Their tactile experiences are highly important during the first five years of life. It is during this time, that creativity is at its highest.

How that creativity manifests depends largely upon their role models who show and encourage physical activities with an understanding that it is fun to learn and play, rather than to emphasize the importance of mastery.

Hero Children will not enjoy being pushed into doing something unless he/she is in the mood to experiment. Once they try something, they are likely to master it quickly according to their own level of satisfaction, which maybe simply be to enjoy a minimal trial that only provides an immediate taste of a creative skill. At this point in development, the Hero Child does not understand improvement, but rather is focused on productivity. One painting is as good as another. None are prized. They are shown and then simply discarded. It could be that many scribbled drawings are done in one brief session.

The Star Children will only focus on his/her perception within the imminent moment. A simple drawing of self at the table with mommy looking on, or the teacher in the classroom will be very important and will be shown to anyone who is interested. It is important for them to be complimented and his/her point of view about how they see themselves in relation to the picture be understood. They will want their artwork put on the wall so they can see it often. They will also want to display and talk often about other creative pursuits.

The Indigo Children have an absolute need to express self through many creative pursuits and will consistently request new information in order to improve their abilities. They are capable of focusing and sitting for long periods of time. Everything seems to have a pattern to them. A boy will arrange his toy cars in line. A girl will place her dolls in special places that look balanced and safe. Their discovery of art and music mean a lot to them. Without these abilities, these children would find life meaningless. They will describe their world, as they see it, through the arts to express themselves. Art can also include the sciences where the study of ecology or mathematical problems seems to be beautiful.

The Crystal Children have no need to study artistic talents. Their motivation is simply to touch, explore and feel everything they see. Any creative skill evolves later, when they begin to understand that there are patterns to life and ways to arrange things which please the eye. The most likely attraction to keep them busy for a short while is to play musical instruments, develop gymnastic talents or study outdoor activities that allow exploration of the Earth. Whichever talents arise, these children should be encouraged and praised in order to prevent loss of the awareness of the first contact with their creativity. Repetition is important for Crystal Children to encourage them to know there is more to their creative skills.

The Liquid Crystal Child is a child of the future when it comes to creativity. Their constant demand to try everything will lead them into a variety of different talents and skills at a very early age. Each child has a very natural way of absorbing information which they use early to express their personality. Their choice in exact skills and talents will depend primarily on their *Soul Structure*'s coding and the inherited talents of their parents. In no time at all, they show a quiet genius self in some way. They might play the piano very well by age four, or do mathematic problems in the blink of an eye. To many, they may seem overwhelming as they show their skills.

As parents of any of the children, it is important to give them all a variety of experiences. Understandably, lessons in arts and sciences can cost a lot of money, but not all children will desire to have a consistent lesson in one subject, so never think about the expense as a long term one. Offer your child a variety of experiences, one at a time and allow him/her to tell you what they like best. Then focus on that one thing until it becomes boring. At that time, introduce another study. Over a few years, you will see very clearly which of the skills and talents your child wishes to pursue.

It is important to not impose your own desires on your child. Your standards are going to be very different from theirs. They may love to play the piano spontaneously and hate to study, whereas your ideas of when and how to practice maybe important to you. In time, they may come to know more than you do about their interests as they develop their own natural talents and skills without pressure. Be sure to support and encourage them in as many ways as you can. Overcome any parental insecurity you might have. It is okay for your child to outwit you and outsmart you sometimes. Never make a child feel inferior. Encourage them to understand that they will improve with practice and are free to choose when to do so.

Being creative with your child ensures that the left and right hemispheres of the brain become interactive at this early age. A lack of creative experience will lead to a more literal and intellectual perception of life that may cause your child to never have any fun in life. Still later, they could develop psychotic tendencies with manic depression or obsessive compulsive behavior. Too much physical activity can lead to ADDH symptoms or, later in life, an emotional imbalance resulting in symptoms such as Tourette's Syndrome, Bi-polar Disorder or diseases such as Multiple Sclerosis.

A child may choose to be left or right handed. Some are ambidextrous. Left handed children are mostly driven by the right side of their brain. They experience life and then put it into their feminine self (yin), which evolves them into spiritual awareness and growth at a late age. Usually, during some crisis in life, they transfer this insight into their practical life by adapting their left brain to demonstrate outwardly what they know. These children are often controlled by circumstances and the people that endorse them.

Children who develop the right hand as the dominant one are most likely to focus on things that surround them. They intellectually process and interact with what occurs around them, and then, through their masculine self (yang) will focus on understanding their practical life and its purpose. In this way, they will come to understand, day by day, the purpose of their experiences, which over time, evolve into psychic and spiritual awareness as time passes. These children become controllers and frequently take charge of others.

Those children who are able to use either hand in any way, are using both sides of the brain in an interactive way that will enable them to be able to experience and understand both spiritual and physical activities while processing everything intellectually. Their talents and skills can be many, as will be their psychic skills. Most of these children are pioneers in the development of the future Earth and the people who live on it. Some of these children may be autistic or have physical and mental problems, but ultimately, they will succeed in bringing about a change, simply by their presence which will influence the people of the world.

All these children will be drawn to various ways of communication. They will create movies, instruments that improve technology, ways to develop and rejuvenate the land and find new ways for our countries around the world to share. The future is in their creative hands.

Their love of beauty and their need to preserve anything that is unique will lead them to design and improve those things that fail us now. Their imagination and joy of exploration in new ventures will ignite a passion never seen before in Mankind. The common thread of a Spiritual *Goal of Acceptance,* leading to Ascension and evolution will lead the entire population of the world to inhabit this planet with better management skills that preserve and increase the world's fruitfulness.

Creative skills will lead to our development of higher technology and the use of energy in what maybe seem now as fantasy. There is more in Heaven that is yet to be manifested on Earth. The Way of The Oneness is to create more. In this way, God is manifested in very many more ways that ultimately will test us and Itself to know ourselves more fully.

Until we understand that God is in each of one of us and that we can accept the many facets of God manifested within us as our true self, there must be events that will bring us all together as one united force. To reach a state of harmony on Earth, there will be a continued increase in our development and our understanding of the Universe. What we know now is very little. Our perceptions and the way our brain works is an enigma, but little by little, the human species will learn more about our make-up and the marvelous nature of Mankind.

It is the very essence of creativity that spurns us to discover more about our nature. These five kinds of children will ensure that we develop a better world to live in.

Ways To Encourage Skills And Talents.

- As soon as your child is able to look at you and smile, tickle him/her and play peek-a-boo games.

- Touch your child all over as he learns to awaken to touch with tickles and develop laughter in game fashion.

- By the time your child is sitting, play music while rocking, jumping exercises are active.

- During his/her sleep time, play music, tell stories and sing to him/her every night.

- By 10 months, encourage your child to play with toys that make noises in a variety of ways. Constantly smile and praise no matter how awful the noise may be.

- By age 1 year, allow your child to scribble on any surface with washable crayons, during which time, teach them the right and wrong places to draw. Encourage a continuance in art perceptions.

- Introduce games that allow your child to learn letters and numbers that involve playing physical games. Play these games on a daily basis.

- Introduce musical toys and exercise play toys, such as jumping toys and slides. Let your child feel the joy of movement in dance.

- By two years old, encourage your child to go to play-school, making sure that there are plenty of artistic and creative pursuits to follow. During this time, the act of sharing with other children will develop their own need to improve their abilities.

- Discourage any competitive/judgmental ideas by showing them how easy and okay it is not to be the best, but rather to try their best in everything they do.

- Help your child develop their body by doing physical exercises at the park or in a class. Tumbles and flips are a lesson in balance.

- Discuss everything your child presents to you, even if it is paranormal. Developing their mind and their point-of-view is vital before they reach the age of three.
- Expand their internal imaging with descriptive discussions and stories that involve imagination.
- Introduce play-dough and other products that allow them to develop tactile senses. By covering their bodies with stuff they will develop a sense of what they feel comfortable with and how to use products well.
- Encourage games that allow your child to use their senses separately, such as "Blind Man's Bluff", a game where the eyes are covered and the child has to feel and guess what it is they feel, or who they are touching.
- Cover your child's ears with a headset and allow them to watch friends while not being able to hear what is said. Enlighten them about how precious it is to hear and to listen to the sounds around them.
- There are many creative pursuits that you can find at home, such as helping make cookies, putting their toys away, etc. All practical things should be done with a sense of fun and pleasure.
- Encourage your child to hop, skip, tumble and dance. Movement and rhythm along with vocal singing is a must for all children. In this way they expand their lungs and get plenty of oxygen.
- Encourage older siblings to share and demonstrate their talents to this younger child.

Things You Should Never Do With Your Child That Would Hinder Creativity!

- From birth, refrain from leaving your child too long once they awake.

- Do not leave them with a dirty diaper for too long. This discomfort can cause a block in the way they feel things on their body.
- Try to not make sudden loud noises that will frighten them. Do not yell at them, as they are too young to understand your mood.
- Avoid take things away from them when they are happy playing.
- Do not leave your child unattended for any reason. Take them to the bathroom with you. They learn about different smells and will need to explore them.
- Never punish your child with slaps or reprimands without "show and tell" first.
- Do not force your child to play.
- Do not push them to experience something they are not sure of.
- Absolutely avoid enforcing your type of music on them. Let them play and discover with a radio or listen to a variety of things.
- Avoid adult TV programs – they are watching and listening and will absorb fear from these programs. They have no separation from fantasy and reality.
- Never leave your child with someone you are not comfortable with yourself.
- Do not let your child play with dangerous toys that could cut them or that they could swallow.
- Do not leave your young child with an older child to care for them.

I am sure your intuitive self will kick in often as you interact with your child/children. At all times, it is important to focus on the safety of your child and their abilities to grow and learn at a faster rate than you would expect. You see, your growth was at a slower rate, because you were born during a time when life was slower. So, do not make the mistake of expecting your child to turn out just like you.

Chapter Twenty One

Intelligence

Society has often defined intelligence as the ability to understand. Since education demands that each child learn to read and write, it is often over emphasized to the point of destruction. So, let us think of a child who is born into a society where there is no form of education. In this particular example, this child has to rely on their observation to understand the meaning of everything that happens around them. They are forced to look, watch and copy. In so doing, they begin to understand, for example, that many coins equal more money than only one. They learn to exchange objects for other objects. In the course of time, they learn to establish their own idea of what is valuable to them. While one child may enjoy a piece of chewing gum, another may prefer to exchange it for a crayon. So, what is it that causes a child to choose something and reject another? Quite simply, it is the desire to survive within the realms of their accepted comfort zone. Going outside the box does not exist unless someone teaches them that there is an alternative.

If we go back in time, when education was only for the wealthy, children all over this planet were surviving without reading and writing. They had no crayons or paints. They simply improvised with clay, berries and the like. Their creative works were often borne out of work schedules. During a hard day's labor, children evolved their ideas into hopes and wishes. Those dreams of a better future inspired them to think of ways to earn money, own property and build a comfort zone around them that would probably be protected with their life when invaded by marauders.

It is the nature of Mankind to develop on all levels of awareness from a physical, mental and emotional point of view in order to access a spiritual one. From the beginning of time, we have become taller, emotionally deeper, spiritually more aware and infinitely more intelligent. Now, in our time, we have developed brains that can logically calculate dimensions of existence that until now have never been thought of before. We wonder about the stars, Universe, God, and then explore them! We think about mathematics, explore calculus and speculate the compounds of galaxies far away. Science has taken us to the edge of brilliance. We are currently developing our brains at an incredibly fast rate more than ever before. We are on our pathway to developing super-minds that will eventually be considered the norm.

Children born in this century are demonstrating their amazing abilities to understand the applications of computer programs of all sorts. They are able to learn many languages at once and to flip from one to another without needing to translate. They can absorb information at a much faster rate than those of fifty years ago. These children ask many questions and expect intelligent answers from their elders. If no answer is forthcoming, they will search until they find one.

Since it is impossible to know how a child will develop their intelligence and subsequent education on a variety of subjects, one can only feed outside information into them with a hope that something will spark them to remember. At the turn of the last century, most educational institutes insisted on repetition and punishment to implant a variety of boring subjects, such as Latin, Greek along with Geography and History. Since very few children were interested in laws and reformation, few remembered their studies after leaving school.

In this century, children have a choice as to what they desire to learn. They quickly make it clear to their parents about what subjects they like to study. Some children are more inclined to study facts, while others like to assimilate information through visual and emotional responses. The combination of intellectual assimilation, visual effects and emotional responses can either make or break a child's will. Since there are many enticing educational computer programs for learning, most children are developing their ability to learn quickly. Some children do find this type of learning difficult. Those children are the physical type with high emotional desire for tactile support. They learn best by copying, rather than watching from the side lines.

At this point, we must begin to be aware of a child's coding from within his/her Spirit. Their *Soul Structure* will stimulate their brain to either learn through feeling or intellect. As their personality develops, so do their mindsets. By the time they are three years old, they will have shown their preferences. So, take a look at your child and determine if they are leaning more to practical activities that involve interaction, or if they prefer to sit alone and amuse self. Neither way is best. Each way must be nurtured while also directing your child to do the opposite. In this way, they will develop a balanced attitude about how they see self and how well they learn.

Memory is a product of emotional acceptance. If a child likes something, they will master it quickly. If they dislike learning, they will develop antagonistic traits that will force them to express negativity. Negativity is an emotional expression of failure and fear of non-acceptance. Since no one wants to be a failure, this negative emotion will stimulate them to want to try harder to achieve acceptance. In this way, a child learns to study harder.

Discipline must be instilled. A young child is capable of exploring anything that moves. Since there is no awareness of time, place and state, your child will need to be educated to understand that there are moral and ethical codes that need to be followed. I once caught my children gleefully stamping on an ants nest. I pointed out to them, that the more they disturbed the nest, the more ants would appear and work twice as hard to rebuild what they had destroyed. After staring at the swarming ants, they realized that they were climbing all over their legs. Once they saw this, they each became uncomfortable and immediately saw their own discomfort in having ants invade their body. Through this, they were able to see the parallels from the ant's point of view. Their space had been invaded and disturbed. This was a good lesson for my elder sons. Later, when their younger brothers were around an ant's nest, I heard them diligently explaining why these two children should leave the ants alone. It was a wonderful moment to see older children educating younger ones.

It is always important to acknowledge a child's intellect and assist them to share their new found knowledge. Every family has something to learn. Perhaps the teachers give your child/children something new which you may not know. We are all pupils and teachers to one another. The application of discipline will help your child organize their mind, prioritize and categorize those things most important to them. These early ideas will lay the foundation stone upon which they will build the rest of their life.

While their point of view may change from day to day, their ability to process their experiences will inevitably develop their wisdom, which eventually will open them up to a higher consciousness, where their spirit mind will break through their conscious mind, giving them a completely new point of view about him/herself as the years go by.

The Hero Child will often insist that they be allowed to study the things they feel important to their future. They may often go out of their way to master a skill or talent. If playing a computer game, they will desire to always be the winner. Each time they lose, they will be frustrated with their results and are likely to develop a sense of inadequacy. It is important to instill into him/her that failure leads to great successes later, after one continually practices and improves. Some Hero Children may develop strong desires to take charge of younger siblings and will become little mothers or fathers while showing their knowledge by directing educational subjects, such as story telling which gives them a sense of control.

The Star Child will most likely wish to study artistic games, make paper chains and share creative stories with family and friends. They will always ask questions and ask why something is the way it is. When given an answer, they are apt to question the answer. They rarely accept one explanation. These children will be drawn to fairy stories that teach a lesson, or movies that show an emotional experience that will set them thinking. Many of these children love to watch such shows as Sesame Street, Teletubbies and the like. They will squeal with delight and then pretend they are one of them. In this way, they learn.

The Indigo Child quite simply loves to hear fairy stories that teach about good versus bad. They learn quickly to identify with what is good, and frequently get very upset when something goes wrong. If another child interrupts their learning schedule, they become very discontent. If the teacher does not give them undivided attention, they will do something outside of the norm to get it. Learning for these kinds of children must be pleasurable and sensational. The ordinary is boring. Simple things like counting must be made into a game where they feel they are successful. Failure to be the center of attention while learning can easily cause this child to become withdrawn and unsociable. They love to dance and jump and scream their delight when success occurs.

The Crystal Child hates to be forced to learn. Their intellectual pursuits must be short and to the point. They learn quickly, but soon become bored. So, to hold this child's attention, one must allow them to choose the time and ways of learning. Bricks, sticks, cars and dolls will play an important role in their learning, as will musical instruments or any computer game that has sound effects. They will play to gain the points and hear the sounds, but not to beat the scores of others. They love to watch interesting features, such as water falling, stones tapping together to make different sounds. It is up to you to become the entertainer in order to get your child to learn.

The Liquid Crystal Child requires a bit of all the above. They want things to hold, play with and discard when they master whatever it is they wish to do within the spontaneous moment. Toys that make sounds such as speaking to them, displaying images or making sounds will hold their attention for a long time. They will keep pressing buttons to hear, see and feel until they can completely remember the sequences and the content of the way the toy behaves. At that point, they will consider it useless. Since they master everything very quickly, they will desire new toys often. Computer work will hold their interest provided the subject has room for expansion as their brain absorbs more information. These children have the potential to become wiz kids.

The more interactive you become with your child, the greater their learning curve will be. Since one can never know what they are thinking, it is best to always acknowledge every conversation they start. Listen well, ask questions and provide answers that will satisfy them completely. If they hound you with questions, be patient. If you do not know, admit it. Then take your child with you and explore together. By showing your child that you do not have all the answers, he/she will understand that even adults learn and so this will encourage them to accept that they too will always be learning. Visits to the local library are a must; let your child choose the books they want to read with you.

Parents who leave education to teachers and who also make a point of not becoming involved in their children's homework and play, often unwittingly help develop a lack of interest in learning. When a child is older, their work schedule will seem unfulfilling and consequently, they will fail in many subjects and leave school with barely any education.

Since no one wants their child to be dissatisfied, it is important to educate them to understand the joy of learning and to instill the importance of expressing what is learned. Any exploratory journey that takes them into unknown aspects of their experiences will be exciting and extremely interesting for awhile. Once they have obtained some information, they will move on to seek something else to learn. Do not be disappointed if your child has no interest in the things you show them. They have to find their own interests and may well bring your attention to something new for you. Some children may like construction kits, while others may love pulling this apart. Either way, they are learning how things work.

Always remember that the emotions created by a child ensure that they retain knowledge throughout life. The more emotional joy there is in learning, the more information they will recall when stimulated. If you want your child to be brilliant, then encourage all their positive emotions that provide immediate excitement with challenges to discuss and share openly. Do not over-crowd the mind with logic or try to insist that they master many things at once. If your child is smart and able to focus on more than one thing at a time, teach them to enjoy each thing, but point out to them that there is plenty of time to spend on one thing at a time. Many children try to do many things at once and end up confused later in life and unable to make mental choices. Subsequently, they spend their life in confusion.

When conversing with your child, sit at eye level with them. Maintain plenty of eye contact accompanied by lots of up and joyful voice intonations. Remember that they are watching your every move. Growing up is supposed to be happy. Learning is supposed to be fun. Never punish your child for not learning. Simply point out that if they miss this lesson, they will need to do it again later. Children usually come around to learning very quickly. They want to be as good as their friends and feel capable in holding their own at any given moment.

Self-Esteem

During the first two years of your child's life, they are learning to not only love all your family, but more importantly to love self and their quality of life. Your child will need to be encouraged to look at him/herself in the mirror and to like what they see. Try to compliment them on their appearance and point out how nice they look in their clothes, especially when holding something they have just created. This visual picture, together with the sound of your voice will always stay in their mind as a primary belief.

Self-esteem is the ability to enjoy being in one's own space and to love everything about one's self on an outward appearance as well as inwardly. So, allow your child to choose what they want to wear each day by opening the draws and cupboards and displaying all those clothes. They will know what they want to wear, even if you do not agree; let them wear those items of choice for awhile. Later you can ask them to change for another event that will encourage them to like to change and adapt. Have a mirror hanging in their bedroom.

Compliment them on the colors they choose and emphasis how colors make them look -- smart, pretty, happy etc. Encourage smiles and add fun to dressing up by having days when they are allowed to put on mum's or dad's clothing to introduce the idea that one day they can grow big, and tall and strong and be just like you both.

Point out how his/her parents and family members all have nice ideas and like to dress well. Sloppy clothing will lead to a sloppy idea of how they appear, so dress you up when going out.

Even if you child has a deformity, or affliction, make sure you constantly tell them how beautiful they are and how wonderfully nice and calm they are too. A child loves to know that they look and feel good. If they have grumpy days because of illness, simply stand there in their room with the mirror nearby and say something like, "Oh you do look miserable today because you are sick/angry/frustrated etc. But, let's put on something different and then you will feel better." Remember that tactile senses are calming and nurturing. Hugs of course will follow, but your baby may be irritable and seem unmanageable. Simply let them take things on and off and just keep saying how he/she looks nice and different etc. They will eventually calm down and fall asleep, having exhausted themselves during this uncomfortable time.

Self-worth

As your child learns, many things are skipped over. Counting is one of the first things a child must learn; 2 shoes, 1 coat, 2 gloves etc. Every stage of learning is important and so praise is equally important. So many adults grow up taking these early days for granted, having lost the joy in everyday activities such as the time it takes to get dressed, bathed or prepare for work. So, tell your child frequently how clever they are to be able to do the skills they have acquired in just living each day. These activities form the foundation of teaching the brain to appreciate all the things they are capable of doing. Later in life, they will transform this part of their education into realizing that everything they do is important and therefore, will lead to something else that is special. When obstacles arise, they will be more than ready to cope with them without becoming depressed. When a child learns to value everything they do, they develop a good ego level along with a decent belief in what they do during each day. In this way, they will become very engaging adults who never cease to appreciate each day that passes.

Self-value

You child must learn the value of time and space. Though very young, their internal time clock is ticking. They know when they are hungry, when to sleep and quickly learn the difference between night and day. It has been said that some adults are best a night because they were born during the dark hours, and vice verse for those born during the day. Perhaps this is true! But, in my observations with clients and patients, I have found that much of their psychological and emotional problems in dealing with time and space do come from a lack of support. Most adults rush around trying to get too much done in a short time.

When you child is around 2 years old, he/she will become aware of you and your time schedule. If you are rushing off to kindergarten, so that you can get to work on time, then your child will learn to hurry and can quite easily become frustrated and agitated. In order to overcome this type of event, try to establish a time structure for you and your family that is amicable.

During the daylight, try to get your practical chores done while you child is sleeping. Then when they awake, you can give them your time to help them learn about time. Give them a routine and praise them for having done simple chores like, put dishes in the dishwasher with you, or to put toys away as they finish playing with them, rather than pulling everything out at once and then running off and leaving them for you to pick up. These simple rules will be followed and accepted at this young age, and will lay a solid foundation for organizing their time and finding ways to enjoy their life without getting stressed when under pressure.

Many adults, who cave in when life does push against their routines, will flake out unless they have been taught discipline in the early years. Discipline is something a child must learn as soon as possible as it will be the ultimate mental foundation stone upon which to build future structures later in their life. Being organized and on time gives a good sense of success when measured up against the talents and skills that your child uses.

These three aspects of awareness will provide your child with a wonderful perception of their life and the people who share it with them. As the years pass, they will fill their working life with productivity that leads them towards a very successful career, while ensuring they develop routines that include hobbies, fun things to do and most importantly time out to rest and rejuvenate for things to come.

Chapter Twenty Two

Maternal Love

I have shared my insights on the rearing of a child. To finalize this journey with me, it is important to end with the importance of maternal love.

The moment a child is conceived, even if it is through tragic events, there occurs a spiritual bond. A child cannot be born unless this bonding has been agreed upon, either in the Spirit World before embodiment or later during a woman's life. Whatever the story, there is a need for love. Every person on this planet needs love. In fact we all crave it.

As soon as a child is born, he/she is on a quest to find that love. Whether it comes in the form of hugs or discipline, even cruelty or sacrifice, there is always some exchange of emotions. All emotions stem from a deep seated spiritual programming that lies in every fragment (Spirit Being) within the Universal Consciousness of The Oneness. Every fragment craves to belong to God and to every other fragment. So, whether your child is with you now, or has been with you in a past life, there is a bond that unites you. Your spirit selves are tied together forever.

No matter what your lesson is, something deep inside you will push you to put your child first. Only in extreme pain, can you even remotely face the possibility of not having your child in your life. Even then, should you give up your child; you will always wonder where he/she is and what they are doing?

This spiritual bond can never be broken on Earth, though many may try to put mental and emotional distance between them. Your spirits are tied in ways that extend your lessons into the collective consciousness of your Soul Groups. Each Soul Group is in some way connected to other Soul Groups and so on, so in an earthly life, your interactions have far greater lessons than you realize. Spirit Guides and angels who watch over you are a part of your lesson. They learn through you and share with you. In this way, we are always tied together and through our interactions in any dimension, will always extend our emotional understanding of unconditional love to one another.

In life after life, we will explore the many facets of love. During lives where there appears to be no love, there is always someone, somewhere who is caring about that person in some small way. No one is completely alone. No one is ever lost or forgotten. No one is cast out or eliminated from God's presence. We are all part of God and are embraced by the Creator's love. In this way we are protected from our own foolishness and encouraged to grow spiritually. God's Love always calls us back in an ever ascending spiral of spiritual growth that will always have an everlasting effect on The Creator. God learns through our journeys.

Christians think of Mary, the Mother of Jesus; Buddhists think of Quan Yin as the perfect woman who loves without judgment. This model of the perfect female can be found in every religion around the world. Since the dawn of time, Mankind has recognized within him/her self that this thing called emotion has a purpose.

That purpose is to explore the many facets of creation in all its forms. Ultimately, it is our desire to know every fragment in all its ways, whether on Earth or in the Heavens and in so doing to know ourselves. As a perfect mother, everyone learns to accept anything anyone does no matter how bad. To forgive is not to forget, but rather to understand. Understanding is simply the ability to be able to say that you have experienced something to the fullest of your capabilities. You can only identify sorrow or happiness by your own experiences. Only then can you say that you have truly found your emotions and have come to 'know' who you are and to be able to see yourself in those you love.

Each one of us expects to be involved with others. As we interact we expect to be forgiven for our mistakes, respected for our efforts and followed by our examples of truth. By living a life that is fulfilled, each one of us can become a role model for others to copy. In becoming a good mother, you are helping in the evolution of all spirits, their Soul Groups and our ascension into the heart of God.

The Eternal Mother is in all of us. We have all incarnated in both female and male forms, and in every case, the issue of maternal love has always been present. Our spiritual journey is to learn to accept that while a mother may like to be perfect, it is in the very existence of imperfection that truth is revealed. The Eternal Mother is both good and bad. She sees all, understands all, and accepts all. She wishes to change nothing, knowing that in all things there is something of value to love.

I hope that in days where you seem to be pressured by your child, while the world around your squeezes in on you, that you will find quiet moments to tap into your own Eternal Mother self and find the essence of your spirit to capture the purity of your ***true soul ascended self***. Without a doubt you will be felt by your child, who will respond with a great deal of love for you.

Epilogue

After Thoughts

No matter what type of child you have, whether more intellectual or emotional; Hero, Star, Indigo, Crystal or Liquid Crystal, all children need to be helped to develop a world in which they can appreciate him/herself on a daily basis as well as to appreciate this planet called Earth.

Always take time out to check your child's emotional and mental state. Avoid creating dysfunctional family issues. Spend time explaining and nurturing, no matter how hard the times are. At this time of writing, thousands of children are undernourished, neglected and abused though physical or verbal demoralizing words and actions. In Third World countries, children are forced to provide for themselves by begging, stealing and mischievous activities that later lead to serious crimes, even to the point of murder. These children need comfort, love and support. They need education and people to help model their lives.

In the overview, all children have chosen lives that will aid their spirit self to learn through hard lessons. But fortunately, some of those hard lessons will lead them to become great leaders in a time when the world is ready to listen, unite and make this place a better in which to live.

Those children who are born in comfort need their parents to teach them about those who are suffering with lack of moral, ethical and financial support. Each child should learn to respect their elders, honor their country and the systems that sustain their lives and above all else, to be taught about their psychic senses and to use them in their daily life to develop acute awareness of one another.

In this technological age, simple graces, along with fine conversation has been lost as a result of the television/video market as well as cell phones and electronic games. Children need physical interaction with other people. They need conversations that stimulate their minds to move beyond their environment. They need games, sports, and theatrical events to open up their imagination. They need the arts to explore their capabilities and time with animals to know and appreciate other forms of life.

Travel has become more economical and faster and we should encourage every child to visit at least one other country to see how children of other nations live. This simple activity will open the minds of children to face truths and realities that might otherwise never have become known to them. In this way they will learn to appreciate their families and the things they have.

Every child should be acquainted with religions, philosophies and theologies from all kinds of cultures; be it in the heart of Africa or in the middle of Rome. Every thought about self and the enormity of creation is a topic for discussion, as is death and our subsequent journey into the Spirit World - The Hereafter etc. Everyone has a fear of dying, but the children of our future already know that they continue after this life and are not afraid to face the passing away from mortal flesh. However, they are afraid to be spiritually left behind and not to receive unconditional love from God. It is a new generation with a new focus on love and it is the important lesson of this time.

Every child should be encouraged to take psychic development and natural healing courses that will aid them in understanding the very essence of their nature and the many alternative medicines. Natural remedies were used by our forefathers with great success. We can teach our children to use those remedies again wherever possible. Right now, a child should be brought into the understanding of the joy of nature and the need to protect this Earth.

Though important, education is not always about reading and writing. It is much more important to teach a child how to survive in a way that will bring happiness to him/herself and others without invading the space of anyone which, when done, often causes suffering. Frequently children become bullies, taunting the weaker ones. Judgments are thrown around like stones. A great deal of harm is done to a child by his/her peers. Teach your child to love and honor all their school friends and to be supportive in any way possible. Ultimately, children will learn and find a new way to build a social structure that will define their place in society. The ways of the Piscean Age are past; we must embrace the Aquarian Age where the focus is on Ascension.

Finally, what is Ascension? It is the act of change from one state to another. People all over the world are seeking change. It takes time for each individual to transform their life, especially while they are carrying the ideas of the Piscean Age. In 50 years from now, no one will think the way they do today! By then, our children's children will have a new way of looking at change. They will see it as exciting and exhilarating and will often meet the challenge of change with great joy.

As the world turns and many changes occur, so each child will turn more and more to their inner encoded Spiritual Consciousness and become extremely aware of their personal connection with The Oneness and with God.

Our entire experience on Earth is a test for every individual to recognize the spiritual Self and to reacquaint themselves with God, The Creator.

Within nothing is something. Or, within something is nothing! Explore, discover and behold!

Appendix

In The Beginning...

In the beginning of time, for the purpose of understanding, a time when God was alone; His/Her energy was intense and dense. God was the first; He/She existed in perfect completion and was extremely satisfied and accepted by Self as 'The All Knowing God.' This was '*The Divine Sage*' in perfect Existence. But, after a long period of time, satisfaction transformed into loneliness and discord. God in His/Her own wisdom felt a need to create and, in friction, caused a deep-seated need to separate and expand self into two parts; one active, the other passive. All that had been whole before was recorded within each of these two parts, but there was a difference: each part was now encoded with a new consciousness, that of dissatisfaction; a desire to separate and explore. The active part of God that was *The Divine Sage* took on a new form now called an *Archetype* – a more meaningful Lower-self called *The Sage*.

Since a desire to explore meant moving further and further apart, the passive part of God felt a need to stay dormant and receptive to feedback from the active part that was to explore and create. In this way, the two halves would still be linked and information integrated as a whole. The passive part of God, the original *Divine Sage*, was open and receptive to new information. Both parts of God were expanding energy within, growing and adapting to all forms of manifestation in any dimension, while always remaining connected and being receptive to all new forms of expanding energy.

Since the original coding of God was now divided, the active part, in replication, divided again and in that moment consciousness was born

This *Supreme Consciousness* gave birth to creativity. Creativity gave birth to a new *Spiritual Archetype - The Artisan*. *The Artisan* now motivated to feel and explore further, separated again and gave birth to empathy in consciousness evolving into a new Spiritual *Archetype - The Priest*, whose awareness of self brought forth compassion. Now that there were three aspects of God in action, *The Sage, Artisan and Priest*, it was necessary for a separation once again to explore further. And so it was that the *Spiritual Archetype – The Priest*, in coping the coding of that which had occurred before, wisely separated and created the new *Spiritual Archetype - The Slave* who would serve the others. The purpose of *The Slave* was to remind the others of their source and their connection with their passive God-self.

However, since God, the passive part of self, was observing, it was noticed that there was an ego arising and so a commune of consciousness evolved yet another *Spiritual Archetype – The Slave* that was given the opportunity to once again separate and manifest yet one more *Spiritual Archetype – The King* who, in all his glory, would manifest the power of the passive God in form. But, as this form expressed itself in separation and supreme glory, the egotistical self became stronger and through a desire to be acknowledged, it separated as was its historical coding from the original separation. In that moment, the *Spiritual Archetype – The Warrior* was created as a support to *The King* to ensure acknowledgement and security in return for support. *The Warrior* while serving *The King*, learned that life in all its forms was exciting, with a desire to be independent. *The Warrior*, in turn, separated itself to gain better understanding of its existence. In that final separation, the *Spiritual Archetype - The Scholar* was created in order to collect information from all that the other active parts of God. In this way, the Higher-self God would know what the Lower-self God was doing.

The task of all the *Spiritual Archetypes* was to be in observation for *The Divine Sage*, who in turn was busy watching Its new Lower God-Self in action. Over eons of time all the seven *Archetypes* were stimulated to perform and find a function. Everything that happened was being fed back to the passive part of God which was constantly re-establishing Itself into a new level of awareness and balance. As everything was and is still being *Divinely Assimilated,* so *The Oneness* was created and continues to expand.

From this new realm of Creation, spiritual, emotional and mental awareness was developed.

In time, God had completely separated Itself into many fragments. Its passive Higher-self side was now in total submission to the power of the active Lower-self side that It had created. This active side took on many forms, and slowly over linear time, following the original coding of separation, subdivided and subdivided, creating new forms, some of which became very physical. Over linear time, a multitude of planets within a vast universe was created.

For us here on Earth, we have come to recognize our association with God and how our life is entwined with entities in many dimensions. So it is on this planet, that we call them Archangels and Devils depending on our perceptions. While the story is fascinating, the lack of physical proof gives us a deep-seated fear of the unknown. Yet, within each of us is a desire to recall what has been forgotten. We long to connect to these Higher God-like fragments of the Creator who are part of us, who in turn make us a part of God. In our journey on Earth, we are learning to find our way back to the Passive Creator that is ready to embrace us as whole once again. To do this, we must incarnate over and over again. Each time we live again, we grow wiser and closer to actively being aware of everything that came from the original *Spiritual Archetypes* encoded within us. Ultimately, we will awaken, unite with all forms and return as God fulfilled.

The Descent of a Soul

Since we are unable to recall or understand the actual separation of God into active and passive parts, we must now look to our own selves for guidance. Each one of us has been spiritually encoded for all time with the energy of one of the *Spiritual Archetypes: Sage, Artisan, Priest, Slave, King, Warrior and Scholar.* This spiritual primordial coding is subtle and never really understood by the conscious mind of humans. However, our spirit self does have recall by attunement to *The Oneness* where all things are known. In the mirror image of God, we do have a passive and active side to our nature. Just like God, Our original existence has been constantly separated into fragments called spirits. This descent into separation has far removed us from the original coding. We have become lost in our Lower-Self causes. Many religions speak of this descent as "The Fall of The Angels." In *The Oneness*, this is simply seen as part of God's spiritual growth and experiment. So, no one was bad. No one was lost. All was as it was. Subsequently, Mankind has mistakenly called spirits "souls." This confusion has cast fear amongst all men. A collection of Spirits together can express one unified Soul experience that belongs to a Collective Soul Group. This is hard to follow and way beyond our awareness, it remains easier to think of ourselves as simply 'eternal.'

When we chose to live on Earth, our spirit must now look at our return journey to God and the act of embracing *The Oneness* again. Each of us will take many hundreds of lives to encounter and work with the 'All knowing-self' while in embodiment. We will also work with other fragments (spirits) that are a part of our Soul Group. A *Soul Group* is the collective consciousness of all the fragments that have been created from the original separation. Those who came from the original *Archetype of The Sage* are also Sages. Those who came from *The Artisan* are Artisans and so on… So, we can say that there are seven types of *Major Soul Groups*.

Each one of the original seven *Spiritual Archetypes* is head of a *Soul Group*. They are often called Archangels to fit in with many of our religious beliefs, though they are truly far beyond this description in the spiritual awareness of God. Over time, each of the Soul Groups has intermixed, so it is very difficult to absolutely isolate any part of the origin of creation or of one's original coding from the source that is God.

We like to think that God is listening to us and answering our prayers. If we ask for help, we expect to get it in the way that we visualize that help. In truth, everything we think, feel and see is understood by God and *The Oneness*, but not everything you ask for is necessarily given. The purpose of receiving is to help you assimilate and ascend back to God. If something you ask for will prevent that journey, then it will not be given. SO, be careful what you ask for!

During the time of separation before each *Archetype* was awakened into action, God was moving through different emotional and mental experiences of Self. Those senses were encoded into the *Spiritual Archetypes*. Each *Spiritual Archetype* is working directly into the personality of God's expression and mental attitude, as well as existences in many forms. Even animals and beings from outer space and other dimensions are involved.

Since we cannot know those other dimensions, it is senseless to try. There are many who guess incorrectly. So, let us stay with the growth of the *Archetypes* in our own world.

Our form and abilities have evolved in fragmentation, by repeated separation, as was the original coding which is now embedded within us. This Spiritual nature pushes us in every life we live to find wholeness and Oneness with God. In our Lower Earthly Forms, we are locked into an existence through which God experiences life. Everything we do is God in action. Whether it is negative or positive, it is a way to cause action to teach a lesson of existence and of self-importance. Whatever we do, there is something of value in it.

Throughout immeasurable, non-linear time, The Higher-God Self has taken forms which could be called Angelic beings that subdivided and created as God did. They each have their own power and their own ways of manifesting on Earth and we as extensions of them, are learning through the coding of our spirit to understand the Soul Coding that is chosen when we live on Earth. That coding is called the *Soul Structure* and varies from life to life. However, the original *Spiritual Archetype* coding has never changed. Your mirror image and coding of one of these *Archetypes* is buried deep within you. The following is a simple explanation:

Lower-Self Spirit Archetypes

A *Sage* will always come into embodiment with a focus on **expression** versus **oration**. This means to learn to listen to the heart instead of the mind. Understanding how the mind dominates the heart leads to a lesson in how to surrender to the heart.

An *Artisan* will always be stimulated to design something into form, whether it is useful or not. In making something, there is a lesson in the way we **create** good/bad experiences that result in teaching self to be flexible and truthful, rather than to follow rigid rules that have been established and are full of **artifice** (cunning). Once again, the lesson is to understand the feelings of the heart. This reflection of God is in all our hearts.

The *Priest* will always be stimulated by outside forces to identify self through emotional acts brought about by **compassion** and enthusiasm. This aspect of God can become judgmental and overly **zealous**. Mental attitudes can dominate and destroy any submission to the passive part of God. In this way, like God, we see understanding.

The *Slave* is always available to be in **service** to any of the Archetypes to the best of its ability, but should it be tied down by dogma, routine and restriction, the *Slave* becomes tired and lost; full of anger and judgment. True service to God is lost. A pattern of **bondage** arises in which the active part of God is fighting its passive part. A slave can split itself in half as God did in the beginning.

The *King* is a very active part of the manifestation of God. In form, this *Archetype* can **master** many aspects of itself, its talents and skills. However, its leadership and example may be in question. This fragmented active part of God can be mighty and **tyrannical**, both forgiving and unforgiving. This open-ended battle within self is a direct reflection of God's own inner battle.

The *Warrior* is yet again an overly active part of God. Whether it is to **persuade** or **coerce** self to grow spiritually, it is all about fighting a battle between the dark and light side of God. So, the question is which part of God is the active part? The Warrior is forever motivated to seek out the answer by integrating with the other *Archetypes*. The *Warrior* must use all skills and talents to involve the other *Archetypes* and to reveal their truths so that information can be directly fed back to the Passive Part of God for review. This *Archetype* will leave no stone unturned or finish no battle until the answer is found.

The *Scholar* will always feel burdened by the other *Archetypes*, for the job of assimilating all their experiences is mighty in itself. Every physical, emotional, mental and spiritual event has to be documented in energy form to be held for all time in *The Oneness,* where all things are seen as being fully *Omnipresent.* The *Scholar* will walk many pathways in an ever-evolving cycle of growth towards *Ascension.* He/she will often be caught between realism and fantasy… knowledge and theory… before new spiritual consciousness will evolve.

These seven *Archetypes* are, in truth, the many personality traits of God. As each of these *Archetypes* has separated itself repeatedly, fragment by fragment separating in like manner, until form is taken, there has been a great deal of discord and disharmony between the active and passive parts of God.

Once forms with egos were manifested, it was used as a way to experience self even further, allowing the personality of each *Archetype* to develop through continual tests. At the same time, those developing personality traits were passed downward to other forms and then usually back to the Passive Part of God. *The Oneness* was still functioning, but energy had become dense and many fragments had become isolated and lost in Universal non-linear Time.

A Spirit's Journey

It became necessary for forms to follow a structure and, in our world, we took on human and animal forms. Much of Earth itself is also a part of the original fragmented *Archetypes*. We do not think about the trees, insects and flowers as a part of our existence, of our own energy. We all live with and accept the idea of separation, (a direct coding from the original separation of God); yet we constantly reach out to find union with God, knowing that somewhere deep inside us is the original coding of Oneness and perfection. We hug a tree and feel connected. We hold our pets and children and swear we have met before.

A fragment, now called a 'spirit,' must incarnate here on Earth and in other realms. A spirit must take a new form time and time again to experience the original separation over and over again. This repeated pattern will continue until every fragment has learned to know that union of all created things is the priority and the norm. At this time of our evolution on Earth, we have just begun to realize that we belong to God and must evolve to become a part of God's Passive Self again. We are driven by our own desire to return in unity.

Long ago, when our spirits took on bodily form, we learned to survive in very restricted ways on Earth. We brought into form our spiritual personalities, passions; wants and needs, along with a desire to be God-like. Buried deeply within each spirit's personality was etched a *Spiritual Goal* that provides a driving purpose toward finding completion. Completion means providing an opportunity to unite again with *The Oneness* and with the Passive God-Self.

Every life we live has God's original coding of a *Universal Soul Goal* which ties us to the Original Archetypes of Creation. We all feel our attachments to 'angelic' beings who seem to connect us with Divine Love and Divine Purpose. However, we all need a personal challenge and choose to work with any one of God's *Spiritual Goals* each time we incarnate. Each *Goal* has a thousand and one aspects of awareness to experience and so many lives must be lived in order to experience them all. Since there are seven *Goals* reflecting the seven *Archetypes*, you should see that it would take forever to know and understand each individual aspect of all seven as well as of all their interactive elements.

Throughout *Universal Time*, individual experiences have been entwined to form a web, known now as *The Oneness* which is continually evolving and is available to us all. But, just like scrolling through the Internet, you can never find enough time or enough ways to understand everything that is there. Therefore, you need coding that will give you pointers in the right direction once you arrive on Earth.

The Development Of A Goal
In The Soul Structure

Each life provides us with a small window of opportunity to personally experience a few aspects of God's creation. Fortunately, it also gives us a great opportunity to observe other aspects of God in the many thousands of people we meet, observe and hear about. Vicariously, we can integrate those experiences with our Past Life experiences and *The Oneness* in which a main theme has been chosen to work upon. In each life, everyone chooses one *Spiritual Goal* as a theme to follow. That *Goal* is then encoded into every cell of the new physical body as it grows inside the womb, and it is also adopted temporarily by the mother who brings this child into the world.

The seven *Goals* are: *Acceptance, Rejection, Growth, Retardation, Dominance, Submission,* and *Stagnation.* The following list shows how these *Goals* are used to develop a personality:-

Acceptance: (Positive side). A spirit has a great desire to exist when using this Goal. It wishes to walk in *bliss* with everything going just right. In body, they seem to know exactly what to say and to do at any given moment. This person will know how to integrate their point of view with others and become a living example of love in action. They always enjoy their talents and skills and they love to teach and share. They feel very connected to God.

On the negative side of *Acceptance*, spiritually, there is a natural tendency to notice all things as incomplete. In embodiment, there is a need to change oneself or others, setting oneself on a pathway of self-destruction. A person may be *ungrateful* for their life's existence and the things they own. They often judge and restrict their experience by performing acts in self-preservation and conservativeness, struggling to survive by keeping things the same. Great fear of failure manifests, creating a fear of God and isolation from others.

Rejection: (Positive side). By using this Goal, a spirit learns that it is possible to live with people who are ungrateful and restricted and to not be affected by them. On Earth, they have the ability to act with *discrimination*; making good choices as well as to help others make their choices from an intellectual point of view. They have an inbuilt sense which stimulates them to separate self from those things that are or seem bad or harmful. In the overview, every spirit carries this fear of rejection and must master it.

On the negative side of *Rejection*, people can often develop a personality that lacks belief in self, talents and skills and is, therefore in denial of any spiritual growth. Focus is generally on what is bad, wrong or uncomfortable. Judgment, rationalization, complaints and illnesses are an excuse to escape the spiritual lessons of Ascension. Nothing ever seems right for them, no matter who is involved, what they are doing, or where they are. They even reject God and their ideas of whom or what God is. Individuals feel isolated and alone, being controlled by *prejudice*.

Growth: (Positive side). This Goal provides a great deal of stimulus for a spirit to find new and exciting things to *comprehend*. On Earth, people with this Goal in their *Soul Structure* are often over-active with too much to do, but for these individuals, there is always an opportunity to know themselves further and to come to know others and God on more intimate levels. By knowing the many aspects of self, a great deal of insight is gained into understanding God, Spirit Guides and others, which develops deep emotions.

On the negative side of *Growth*, individuals find it easy to race ahead with ideas; swamped with emotions that cause an emotional and mental breakdown. This causes a loss of focus. When this occurs, one feels detached from the power of God to manifest more. This results in a deep-seated sense of loss within the physical self along with *confusion*. At that point, the spirit of a person begins to face *Soul Fears* explained in the text. Fear of the unknown overrides everything and panic ensues. There is, however, a tremendous level of growth that occurs through suffering and we ultimately will understand the lessons provided.

Retardation: (Positive side). By dealing with chosen physical, mental, emotional or spiritual imposed restrictions, a spirit can evolve and ascend into

Divine Consciousness to experience a unique epiphany. This Goal can give an individual spirit a chance to understand how it is to be neutral and without judgment while at the same time being conscious of everything that is done by self or others. On Earth, history can be repeated (*atavism*) skipping from generation to generation. Habits and traditions play a strong part in everything. Beliefs are an important part of understanding which is stored by the brain as memory on a cellular level. By suffering in this way, a spirit can begin to understand cause and effect and awaken to the need to find spiritual freedom.

On the negative side, a spirit when lost in *Retardation* can be locked in with habits that initiate fears, phobias and anxieties along with family routines that block spiritual growth on Earth. Emotional and mental depression may set in with various hereditary illnesses that can cause loss of life. Mental and emotional stimuli can cause withdrawal and *atrophy* of the body. Ultimately, in this way, a connection to God is subtly re-established. Self praise and adulation is the lesson here, which in turn is reflected back to God.

Dominance: (Positive side). A spirit usually takes on this Goal to improve their *leadership* qualities. However, leadership is, of itself, an opportunity for anyone to establish their own rules to live by. Whether they become scientists or healers, their way will be their own. Their presence is always strong and their words meaningful. They live to share their wisdom and knowledge with those who seek their advice and support. Their ways, however, can be seen by others as overly strong and controlling or threatening. In the mirror image of God, they are usually right.

The negative side of *Dominance* can be disastrous. Should a spirit be overcome with earthly problems while in embodiment, they will surely develop a state of mind that tends to be *dictatorial*. They want everything done their way and right now. They judge and condemn and flare up emotionally at the least little thing. They long for a quiet life, but find it difficult to let go of the issues that control them. They feel responsible for everything they do, and in this negative way, reflect God's responsibility within to find inner peace. Without support they may crumble and lose their way.

Submission: (Positive Side). When a spirit chooses this Goal they are learning to let go of earthly conditions that create restrictions and limitations. They will spend their entire life caught up in some kind of structure that will test them in many ways. They will learn to go with the flow, adapt and integrate themselves with whatever the majority is doing. If there is something that seems wrong, they will simply step aside and observe or walk away, having learned an important lesson in accepting that God leads them in the right way. They are full of *devotion* to Mankind and God.

On the negative side, when they find themselves in a circumstance without leadership, this spirit will in *Submission*, surrender to anyone or anything that seems to give some semblance of peace. Physically, they will become *subservient* to other and their ways. They will do what they are told even when it hurts them. They will hurt themselves with drugs, alcohol or bad relationships, over and over again. They fear change and will stay in a rut throughout a life. They will ignore their skills and talents and see themselves imperfect until they learn to surrender to God and to receive *Divine Love* and support.

Stagnation: (Positive side). When a spirit chooses to work with this *Goal*, they may choose to live in a body that has some deformity of the mind, or a physical default in the DNA strand and genes. In form, they are without control, often living in a state of *suspension*, waiting for something to happen. Those that love them will work with them to help them get better. In some cases, there can be a complete recovery, which in turn leads them to work for the betterment of Mankind. This individual will start life the hard way, working to find ways to rise above restrictions and limitations of the Body and Mind. God's miracles can happen for them.

On the negative side, these people can be born with an inner fear of trust or of connecting with others. They simply sit and watch the world go by. In time *Stagnation* causes further blockages in the body which usually affects the brain. They lack passion to try anything and have little energy to motivate self (*inertia*) to experience more of the world. Inwardly, however, their spirit is observing and learning by encountering those who take care of them. This kind of life is assimilated in the spirit world at the end of the physical life. Their presence in the caretaker's life is a strong motivational lesson.

When a woman gets pregnant, her own encoding of the *Goal* in her *Soul Structure* will entwine with the *Goal* of her unborn child. Energy joins in unity, *spirit to spirit, mind to mind, heart to heart and soul to soul.* All these aspects have been italicized to emphasize the fact that these are the Higher Selves integrating to become one. This unity is a complete sharing and full acceptance of the lessons, personalities and characters that will be entwined while living together in Oneness.

Since the ways of the Earth are varied, each spirit will develop a Lower-Self physical personality and character that has nothing to do with their Past Lives or their spirit personalities and characters. However, their spirit lessons and deep commitments to *The Oneness and to Ascension and Assimilation* are fully manifested into the body, mind and emotions of each one. Once the birth occurs, this difference begins to show itself immediately.

The Development Of The Soul Structure & The Modes

Because a mother and child must share deeply during a pregnancy, their individual coding, which has been selected prior to embodiment, will begin to function the moment they unite on Earth. Each is encoded with various traits that are called the *Soul Structure.* Simply put, this coding develops a reason to live and also gives the human body and mind a form of understanding in how to act and behave. The *Soul Structure* is God's personality and character in form that is encoded within the DNA of each of us. We can each choose which part of our own personality and character to work with during each life, and subconsciously learn to know that within us is the power to manifest a variety of infinite possibilities, all having been created by God first. Each individual coding is very personal for everyone and is used to benefit their own spiritual evolution. Each life lived is closely entwined with various family members, friends, enemies and strangers. Life by life, day by day, each spirit learns more about the

Archetypes and the *Spiritual Goals* as well as their purpose in God's evolution.

For each incarnated spirit to have its own unique personality there must be coding that will ensure independent genetic development. Each being, man or animal, has a personality that is developed through the integration of *Modes*. Modes are simply a dominant way to control emotions, mental attitude and physical expression. The Modes are *Power, Caution, Passion, Repression, Aggression, Perseverance* and *Observation*. Since I have written a great deal about the coding of the *Soul Structure* in my book **The Rejection Syndrome**, it is only my intention to briefly explore these aspects here.

Whether you have chosen to be intellectual or emotional in this life, you will choose three of these *Modes*. For example: *Power, Repression and Perseverance* could result in a person being full of presence that is totally negative and restricted; or a natural leader who rules with traditional ways and means, ensuring that no one gives up the 'old' ways.

Whatever *Modes* you personally have chosen, there will be one or two that you have in common with your child. Here is an example: Two people, both with *Modes of Aggression,* could be fighting one another all day or alternatively, be involved in some mutual project where harmony is felt in the doing. This 'mirror imaging' is essential in that it allows a bonding within the womb to occur. If it is lacking, there is likely to be no pregnancy unless there is a deeper bond from a Spiritual point of view, which will be in some other part of the coding of the *Soul Structure*.

Some people are born with a variety of differences in their coding from their mother's. This means that there is a life's deeper purpose in God's plan for that individual. Often, mother and child will be driven in some way to live their lives separately. If there is a mutual lesson going on when the pregnancy has ended, mother and child could continue with physical and emotional difficulties that would ultimately drive them apart in preparation for what is to come. Both would endure and grow stronger. Since there are many combinations in the way *Modes* are used, there can be many interactions of emotional, mental, physical and spiritual levels.

The Development Of The Attitudes In The Soul Structure

During pregnancy, a mother will be working with her encoded *Attitude* which may be indicative of her first or second cycle, depending on her own journey of spiritual growth. The coding of the *Attitude* is a very important aspect of character building. It is through this coding that every learned thing is processed. The way you deal with each episode in your life is colored by a primary *Spiritual Attitude*. Your general daily attitude in life is not the same as the encoded spiritual one. The primary focus of any cycle under the influence of the *Spiritual Attitude* is to ensure that your spirit learns. Your encoded *Modes* will back it up as you deepen your personality into maturity.

The *Attitudes* are: *Idealist, Skeptic, Spiritualist, Stoic, Realist, Cynic and Pragmatist*. Though you can see all these aspects in our everyday selves, it is the general overall character of your performance that can be seen to be controlled by one of these encoded *Spiritual Attitudes,* especially the primary *Spiritual Attitude*. The other choices of supportive *Spiritual Attitudes* determine the number of years a person lives with this focus.

For example, a child could be demonstrating a great deal of resistance to touching and petting a dog, even though everyone else is doing it. Their *Primary Spiritual Attitude* – a *Cynic* creates a deep-seated fear of distrust and contradiction. Over the years, the child develops a dislike for dogs and refuses to go near one, having integrated this *Cynic* self with the second cycle of the *Skeptic*. No matter what they read, see, hear or feel, he/she remain *The Skeptic* within themselves, believing that a dog will indeed harm them one day. In this way, the *Primary Spiritual Attitude* of the *Cynic* is teaching them to look for harmony in all things. In this scenario, it would be a hard lesson.

If you spend time watching and listening to someone's activities and conversations, you will be able to clearly see their *Primary Spiritual Attitude* operating. An *Idealist* speaks of what will be better. A *Skeptic* will expect the worst in everything. A *Spiritualist* will talk about their personal lessons and problems. A *Stoic* will avoid discussion, keeping things the same for the sake of peace, while entrapped in fear. A *Realist* will be the ideas person, with lots of guess work going on about what they believe to be true based on history only. The *Cynic* will argue and contradict the truth or vice versa for the sake of being different as an attention-getter. The *Pragmatist* will try to solve problems to the best of their practical ability, insisting their way is the right way.

In the overview, each *Spiritual Attitude* can be borrowed and used on Earth. They are extremely interactive with one another, allowing us all to integrate our experiences and get to know one another on deeper everyday levels, thus encouraging spiritual evolvement. However, it will always be the *Primary Spiritual Attitude* that will always override the others, when necessary, in order to preserve the spiritual lesson. As we die, we face our *Primary Spiritual Attitude*. We balance our mind, body and spirit in preparation for our return to the Spirit World. For example, *The Idealist*, having embraced a life of misery and failure would, upon reflection, see those sad events as the most ideal way to have learned.

Discipline Is Created By The Chief Feature

Since we are not all angels with perfect perceptions of our self and others, we all need a darker side of our coding to help us grow. A *Chief Feature* ensures we motivate our mind and emotions to act. We can all exhibit moods, mental blocks, physical imbalances and spiritual neglect. The seven *Chief Features* are: *Greed, Self-destruction, Arrogance, Self-deprecation, Impatience, Martyrdom and Stubbornness.* So, you can use any one of them at any time during your life. However, one of them is deeply embedded within the *Soul Structure* as the encoded key to all your experiences. For example, you could be loving and giving and very understanding, but as soon as something seems boring, you become *greedy* for something different. These *Chief Features* are like a double edged sword. You can use them for good and get bad and vice versa.

The Spiritual Centers In The Soul Structure

Throughout your life, you will integrate the coding of your *Soul Structure*. Your encoded *Archetype, Modes, Attitudes and Chief Feature* will define the challenges you must experience. Fortunately, each of us has *Spiritual Centers* encoded into every cell of our body that will generally ensure that we each do the right thing. For each of us, this is the best way to learn a lesson. With one exception, these *Spiritual Centers* are in pairs; *Higher Intellectual & Intellectual, Higher Emotional & Emotional, Moving & Sexual*. The exception is the *Instinctive Center* which ensures development of intuition from a survival point of view. A combination of three *Spiritual Centers* will ensure you are on the right path and learning your lessons. You have chosen one for your state of mind, one for your expression of emotions and one for your physical actions. Should you use *Instinctive* too, you will have an awareness that calls for daily alertness.

Whether you have chosen to be psychic, empathetic, sympathetic or simply indifferent comes through these encoded *Spiritual Centers*. The more aware you become of God, the more open and productive your life can be.

If you have a *Higher Intellectual Center*, then you will be exceedingly aware of your ability to learn and motivate yourself to stimulate and assimilate Lower-self emotions. The more you cry and lose your way, the faster you learn to listen to your inspirational mind to find answers.

Should you have chosen A *Higher Emotional Center*, you will find yourself exceedingly creative and emotionally unstable at first. Everything seems to stimulate you to feel people's ways and to harmonize with them through emotional identity. In time you will learn to analyze and spiritually comprehend experiences.

If you have a *Moving Center*, you will direct yourself to move on in order to learn more. If you have a *Sexual Center* you will be driven to stay put and always find an answer to a problem.

Any combination of these *Centers* will help you to tap into your true spirit self. When this happens, epiphanies occur.

Using the *Lower Intellectual and Lower Emotional Centers* will ensure practicality and focus on habits, routines and fixed ways of life. In this way, a sense of being grounded in Earthly realities can develop either a harmonies bond with others, or deepen a sense of separation and loss, depending on the spirit lesson required.

Since we can choose to mix these *Spiritual Centers*, every life lived has a multitude of lessons to be seen, mastered and outgrown. The ultimate desire is to move away from Earthly control and embrace the power of the ascended Spiritual Self and then embrace God and the journey of Ascension.

More is explained in my book **The Rejection Syndrome.** (Available from www.AuthorHouse.com or online at www.amazon.com.

Your child is very special to you and to your life. Your journey together will help you both achieve oneness and a growing inner peace once you see and understand the intricacies of the entwining of your *Soul Structures* together. So much love, peace and growth can be obtained as you share your daily lives. Always remember that God has given you the gift of life and the ability to give life to others. Your manifestation of the quality of life is supreme and glorious. Therefore, treasure it and the things you do. Look forward to the future with hope and expectation of wonderful things to come.

Products & Services

From Sumaris Enterprises

THERAPY KITS

Dr. Margaret's Crystal Acupuncture[sm] Therapy Kit

This amazing set contains 8 crystals and 3 pendulums attractively packaged in a satin purse which can be easily carried in a handbag or pocket. Included in the kit is *The Book of Crystal Acupuncture Diagrams*. This clear and detailed work gives directions on the use of the points and pendulums and also presents valuable information on the Chakras, the Five Bodies and the acupuncture meridians.

Dr. Margaret's Teragram[sm] Therapy Kit

Dispel the Madness with our kit containing one each of Natural, Blue, Violet, Red, Green and Pink Agate plates attractively contained in a satin drawstring pouch. A simple instruction booklet provides directions and tips. As a special bonus, we include a CD by Dr. Margaret Rogers Van Coops with a color meditation and a meditation for Chakra and Five Bodies balancing.

Dr. Margaret's "Core" Teragram[sm] Therapy Kit

Releases Negative History stored in your body's cells. Banishes effects of old issues. Strengthens you to deal with emotional and mental issues. Stimulates more efficient energy flow. Generates inner sense of well-being. Margaret Rogers' *"Core" Teragram[sm] Therapy Kit* contains 3 specially selected Agate slices with a Basic Relaxation Meditation CD featuring the nurturing voice of Dr. Stephen Van Coops. You will quickly enter into a deep alpha state ideal for releasing old emotional habits and irrelevant mind conditioning.

Dr. Margaret's Spiritual Crystal Acupuncture℠ Kit

Five small triangular stones combined with four specially selected pointed stones to refine dense energy and to redirect it according to your desire or need. Kit includes detailed booklet with instructions and diagrams. Spiritual Geometry leads to focus on Spiritual reality beyond your previous physical awareness.

Dr. Margaret's TrinityStone℠ Healing Kit

This unique kit allows auditory memory to stimulate and shift negativity from the body. The five specially selected large open equilateral triangles are used on the Chakras, one at a time, and then all together to erase fear and illusions stored in associations with sounds. Each triangle, when added, will enhance your perception, vision and positive sensations. Included is a booklet with instructions and diagrams and two Serpentine isosceles triangles that are useful in raising the Kundalini and stimulating and harmonizing the Higher and Lower Sacred Centers. When all seven triangles are used together, an awakening may be realized that will result in an explosion of self-confidence.

BOOKS

Breakthrough Therapies:
Crystal Acupuncture℠ & Teragram℠ Therapy

While most people today vaguely realize that the body is a working machine that generates energy, most of us don't understand the way that energy flows, where it goes, and what it does. **Breakthrough Therapies: Crystal Acupuncture & Teragram Therapy** is the product of Dr. Margaret's research with her clients and under medical supervision. Her research has validated the integration of the energies of The Five Bodies. The book reveals how the principles of Oriental Acupuncture, combined with the use of specially cut crystals and semi-precious stones, will unblock energy flow in our Five Bodies and will tone, stimulate and balance the Chi energies. Using natural resonating energy stones and crystals such as, but not limited to Hematite, Jasper, Citrine, Amethyst, Carnelian and Quartz, has opened the door to drug-free, inexpensive solutions to emotional and psychological issues ranging from addictions to depression to stress. Dr. Margaret's powerful, non-invasive healing methods also provide remarkable relief from minor physical ailments like headaches to major illnesses and syndromes such as AIDS, Cancer and Multiple Sclerosis. AuthorHouse Publishing Co.

The Book of Crystal Acupuncture[sm] & Teragram[sm] Therapy Diagrams

Complementary Healing Therapy has taken another step forward with this amazing book illustrating and describing dozens of crystal stones and tools with techniques for effectively treating acute and chronic conditions suffered by humans and animals. From headaches and minor injuries to major complicated illnesses, Dr. Margaret's treatments provide effective non-invasive and inexpensive remedies to put you or your clients back into a state of positive healing. Dr. Margaret's work with her clients has further validated ancient Oriental Acupuncture principles and merged them with exciting, simple methods using crystals to unblock energy flow in your Five Bodies to tone, balance and stimulate your Chi energies. Her research has carried this work into the treatment of pets and even wildlife. AuthorHouse Publishing Co.

The Way to Oneness

This inspiring work delves into the cosmology of multi-dimensional spiritual existence. Beginning with the "Word" as vibrational consciousness, this book takes you on a journey through the principles of creation, separation, the descending and ascending currents, faith, intuition, belief and evolution. The various sub-divisional cosmologies of the seven archetypes and planes of existence are viewed. Also, incarnation, reincarnation and the Akashic Records are explained as an inter-relationship with the deep subconscious and the Chakras. Of particularly unique interest is the principle of soul fragmentation that the book discusses throughout the text. ***The Way to Oneness*** concludes with practical steps and techniques for emotional balancing and relaxation, disciplinary exercises and various other psychic tools such as astrology, numerology, graphology and palmistry. Recommended for all practitioners seeking insight into higher knowledge

-- *James Ravenscroft*, <u>Whole Life Times</u> *March 15. 1990*

The Rejection Syndrome

In our daily lives, all of us experience moments of rejection that create an internal impasse, either by ourselves or by others. My intent is to assist those wishing to be free of those encumbrances brought about by **The Rejection Syndrome**. This is about a pattern of existence that compounds habit, routine and conditioning, leading to limitation, restriction, judgment and competition. Learn about the soul structure and how you can use it to be aware of yourself and to perform to the best of your ability without negativity or rejection. AuthorHouse Publishing Co.

50 Spiritually Powerful Meditations

In the stillness of the mind lies the answer to your purpose. Dr. Margaret has tested all of these meditations herself. By doing each of these meditations, you can find true direction for your life and release fears, pains, restrictions and anger acquired through conditioning. These meditations work! Develop your psychic ability, fine tune your healing skills, mend relationships, empower yourself and much more. This should be a book on everyone's shelf. Jaico Publishing

Pro-Life, Pro-Choice, Pro-Spirit

Spirit's truth is clearly shown through Margaret's own personal experiences. Is everything pre-ordained? The word "abortion" evokes emotions in almost all normally rational minds. Right or wrong? Moral or immoral? Should it be legal or illegal? One of the most burning issues of our time: Advocates of both sides have thrown themselves at each other's faces even to the point of violence. This book is a must read for women who have been, are now or are likely to become pregnant. Without being judgmental, Dr. Margaret provides the wisdom of Master Teachers to assist women to acknowledge, accept and deal with their circumstances. She has crossed the worldly boundaries to discover just what really happens from the point of view of the child-to-be's spirit and spirit Master Teachers. AuthorHouse Publishing Co.

Henry's Secrets

In this gripping mystery, Dr. Margaret Rogers Van Coops, makes her debut into the world of fiction to explore human psyche. Meryl Jones, an African-American single mother has established herself in the world of advertising. Upon the death of her mentor, Henry Wiggins, she is plunged into a stream of events and revelations that turn her world upside down. The story careens into a surprising climax…a web of secrets spills out, changing everyone's lives forever…

Expanding Images

This book fully explains the many aspects of the Psychic Abilities. Through understanding Psychometry, Clairvoyance, Clairaudience and Clairsentience (smell and Taste), with practice on the focus of images, sounds and feelings one can be guided to understand the way to develop these skills and to become a practicing psychic. This book also includes a simple collection of pictorial images of every part of the *OmniCard*tm, where in-depth descriptions of the meanings of these symbols are explained.

FOCUSING TOOL

*The OmniCard*tm

The *OmniCard*tm is the simplest way of doing a psychic reading for yourself, your friends or your clients. This revolutionary tool lets you easily tap into images that apply to the question being asked. This can be a wonderful new adventure in learning and psychic awareness. Simply attune to a question and then let your eyes scan the images until one looms up at you. Visualize the significance of the image and all of the meanings it draws forth. It's like having an entire Tarot deck of cards in one stylistic full-color painting, which evokes vivid, literal and symbolic images. These images are the focal points for you to create a psychic reading that will entertain and amaze your friends and clients. Try closing your eyes and letting your hand move over the *OmniCard*tm. Your fingers may point to some image of interest or significance. There are many ways to interpret the answers to questions. Try experimenting with them and discover how effective you can be. Special bonus! We are including *Expanding Images,* a special book and glossary of symbols with your *OmniCard*tm that will make it even easier for you to match the pictures with their meanings.

EDUCATIONAL TOOLS

Audio Cd's

Dr. Margaret Rogers Van Coops has given many informative and interesting lectures, which are available on audio CD's. She also provides hypnosis and meditation CD's for focus on specific problems, issues and conditions. Please contact Sumaris Enterprises for titles and prices.

Personal Services

Dr. Margaret is available for private consultations and is also available to do recorded readings by mail or over the phone. Call: (928) 453-7974

Sumaris Enterprises
**321 Farallon Dr., Lake Havasu City, AZ 86403, USA.
Website: www.sumariscenter.com
e-mail: drmargaretrvc@gmail.com**

About The Author

Dr. Margaret Rogers Van Coops has been an ordained minister and missionary of the Universal Christ Church (School of Spiritualism) since 1983. She is currently the Director of Education and Treasurer for UCC. Margaret is a Ph.D. specializing in Medical and Clinical Hypnotherapy and Behavioral Sciences. She is also a DCH(IM), a Doctor of Clinical Hypnotherapy and Integrated Medicine. She has practiced successfully in Spain, France, Switzerland, India, Egypt, Japan, England, Mexico and the United States. Her professional affiliations include the Spiritualist Association of Great Britain, the British Astrological and Psychic Society, The International Medical and Dental Hypnosis Association, the International Association of Counselors and Therapists, the International Hypnosis Federation, the Professional Board of Hypnotherapy and The American Counseling Association. Margaret was among the co-founders of the International Psychic Forum and the American Metaphysical Society. Her dynamic lectures and workshops in Japan and the U.S. have led to regular invitations to speak and participate in international events, including Whole Life Expos and Lifeways/BMSE Expos in various American Cities and The Festivals for Mind, Body and spirit in London and Los Angeles.

She is the author of six other metaphysically oriented texts, including *The Way to Oneness, The Rejection Syndrome, 50 Spiritually Powerful Meditations, Pro-Life, Pro-Choice, Pro Spirit, Breakthrough Therapies: Crystal Acupuncture & Teragram Therapy, The Book of Crystal Acupuncture and Teragram Therapy Diagrams and Expanding Images*. She has authored two novels: *Regenesis* and *Henry's Secrets*. Her books have been published in Western and Eastern Europe as well as Russia, China, Mexico and India. Dr. Margaret has written screenplays including *The Regenesis Trilogy, Seeing Blind* and *The Survivor*, and she is negotiating production of several reality TV series treatments. Margaret's TV series, Psychic Chit Chat, has been aired weekly on many public access channels in Southern California and Arizona. The show features Dr. Margaret and her husband, Dr. Stephen Van Coops, also a Metaphysician and collaborator on her works.